CAMBRIDGE STUDIES IN ECONOMIC HISTORY
PUBLISHED WITH THE AID OF THE ELLEN MCARTHUR FUND

GENERAL EDITOR

M. M. POSTAN

Professor of Economic History in the University of Cambridge

THE INFLUENCE OF ENGLAND
ON THE FRENCH AGRONOMES
1750–1789

THE
INFLUENCE OF ENGLAND
ON THE
FRENCH AGRONOMES
1750-1789

BY

ANDRÉ J. BOURDE Ph.D.

Agrégé de l'Université
Lecturer in French History and Institutions
in the University of Manchester

CAMBRIDGE
AT THE UNIVERSITY PRESS
1953

CAMBRIDGE UNIVERSITY PRESS
Cambridge, New York, Melbourne, Madrid, Cape Town,
Singapore, São Paulo, Delhi, Mexico City

Cambridge University Press
The Edinburgh Building, Cambridge CB2 8RU, UK

Published in the United States of America by Cambridge University Press, New York

www.cambridge.org
Information on this title: www.cambridge.org/9781107625372

First published 1953
First paperback edition 2013

A catalogue record for this publication is available from the British Library

ISBN 978-1-107-62537-2 Paperback

Cambridge University Press has no responsibility for the persistence or
accuracy of URLs for external or third-party internet websites referred to in
this publication, and does not guarantee that any content on such websites is,
or will remain, accurate or appropriate.

CONTENTS

CONTENTS

ILLUSTRATIONS

LIST OF ABBREVIATIONS

The following abbreviations are used in the text:

A.H.E.S.: *Annales d'histoire économique et sociale*.

B.P.E.: *Bibliothèque physico-économique*.

R.H.D.E.S.: *Revue d'histoire des doctrines économiques et sociales*.

R.H.E.S.: *Revue d'histoire économique et sociale*.

Jour. Oecon.: *Journal Oeconomique*.

Biographie: *Biographie Universelle* (Michaud).

Les Caractères originaux: *Les Caractères originaux de l'Histoire rurale française*, Marc Bloch.

Cours d'Agriculture: *Cours complet d'Agriculture ou Nouveau Dictionnaire d'Agriculture théorique et pratique, rédigé sur le plan de l'ancien Dictionnaire de l'Abbé Rozier*, M. le Baron de Morogues.

Dictionnaire d'Histoire Naturelle: *Dictionnaire raisonné, universel, d'Histoire Naturelle*, Valmont de Bomare.

Dictionnaire de l'Institut: *Dictionnaire d'Agriculture . . . par les membres de la section d'Agriculture de l'Institut*.

Eléments: *Eléments d'Agriculture*, Duhamel.

Essai: *Essai sur l'Amélioration des terres*, Patullo.

Mémoires d'Agriculture: *Mémoires d'Agriculture . . . publiés par la Société d'Agriculture du Département de la Seine* (1801-19).

Nouveau cours d'Agriculture: *Nouveau cours complet d'Agriculture théorique et pratique, par les membres de la section d'Agriculture de l'Institut*.

Traité: *Traité de la Culture des Terres*, Duhamel.

PREFACE

In the following pages an attempt is made to examine the relations between France and England in the second half of the eighteenth century in the sphere of agricultural literature. I have endeavoured in this way to broaden and investigate a field opened up by Wolters in a section of his *Agrarzustande und Agrarprobleme in Frankreich von 1700 bis 1790*. I venture to hope that the material now available will justify the emphasis thus placed on this aspect of the period. Although the work of the French Agronomes has not been altogether ignored by modern historians, it has not so far been considered as a subject worthy of an independent study. I have done no more than draw attention to it by pursuing the general trend of historical enquiry initiated by the work of such eminent scholars as George Weulersse and Marc Bloch.

There is an important distinction to be made between two parallel movements in eighteenth-century France—that of the Economistes and that of the Agronomes. In fact there was a school of theoretical or, as it was then called, *speculative*, agriculture with an independent life of its own, a school the existence of which can be traced in the extensive economic literature of the time. It is with this speculative movement, isolated from pure political economy, that this work is considered. It should therefore be considered as a study not of economics, but rather, of the history of agricultural technique. Its scope has, indeed, been limited to the study of a particular influence, the strongest one, on this movement; namely, that of English techniques. It might appear at first sight that, had the Agronomes ignored what was going on in England, they would still sooner or later have arrived at the same conclusions. This however, was not the case and their appreciation of the English agricultural revolution certainly spurred them to action and caused them to introduce remarkable innovations in practices which, it is now generally agreed, had been both backward and static. The fillip given by the study of English agricultural technique to French research in that field may be compared, with all due deference to literature, to the impetus which English writings gave to French pre-romanticism.

It is, then, within the domain of theory, or at most within the limits of practical but specialized experiments, that this study should find its place. The date 1789 is therefore an arbitrary one. Naturally, it was, in a sense, thanks to the Revolution that some of the ideals commonly shared by both the Economistes and the Agronomes were realized. But whereas the Revolution strongly affected the country's economic structure, it did not do so as far as theoretical husbandry was concerned. There is very little difference between the *Mémoires de la Société Royale d'Agriculture* of 1785 and the *Mémoires d'Agriculture* of 1802. The chemical discoveries of the beginning of the nineteenth century are, in this particular respect, much more of a revolution than the political Revolution. However, 1789 is convenient, as a date, and certainly not without significance.

The forerunners of the movement up to 1750 have been dealt with briefly. It has been necessary to go fully into the work of Duhamel du Monceau and his debt to Jethro Tull, as well as into the controversy he aroused. The principles of the *nouvelle culture* once explained, it was difficult to avoid all mention of their connection with problems of rural economy. Here the Physiocrats seem to have been supported by the Agronomes, rather than to have determined the Agronomic movement. It is at this point that it becomes necessary to trace fully the influence of England in the various domains of husbandry. Finally an account had to be given on the various means by which the Agronomes acquired this practical knowledge and of some of their achievements in

PLATE I

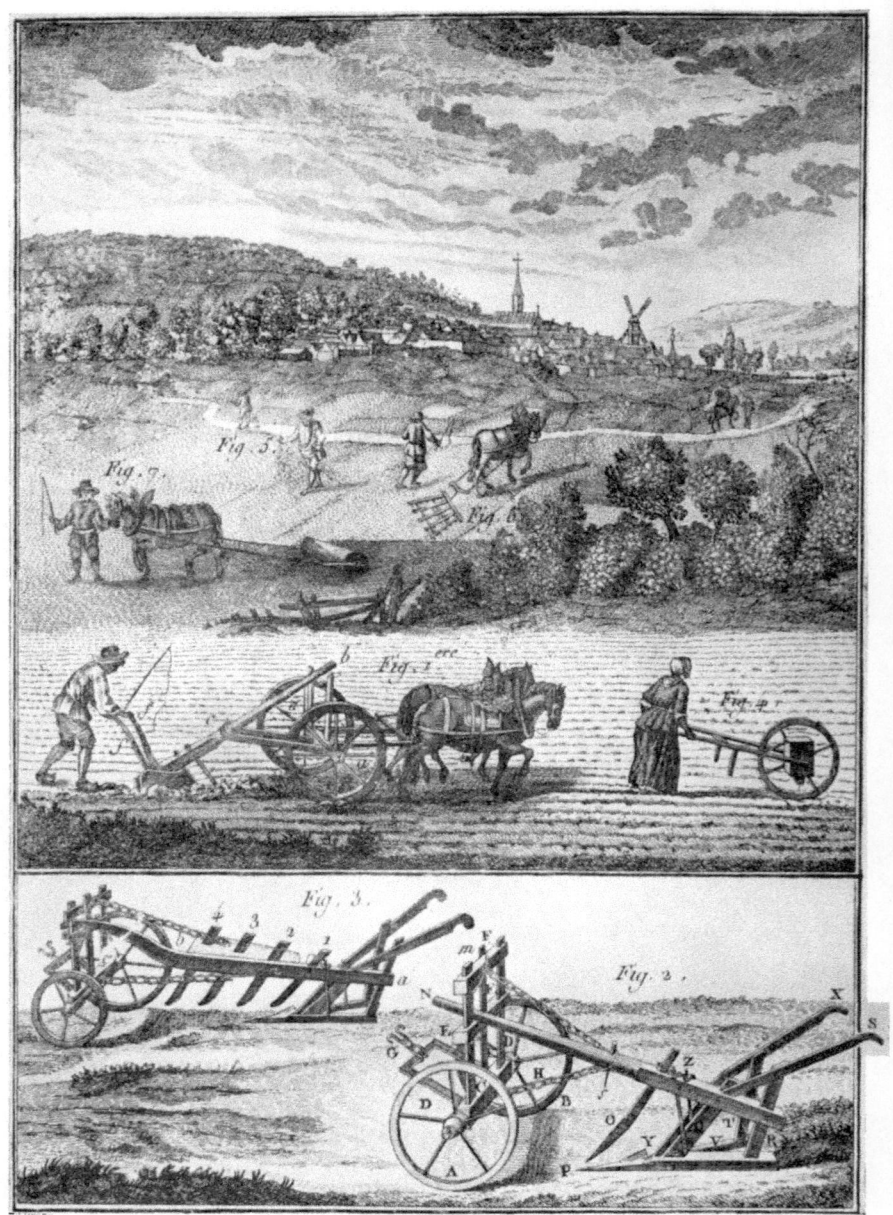

AGRICULTURE: PLANCHES (*Encyclopédie Méthodique*)
The picture shows specimens of improved ploughs (particularly Tull's four-coultered plough, after Duhamel's *Traité*). Notice the use in the fields of the roller, the harrow and Abbé Soumille's drill (*semoir à bras*) and the exclusive use of horses in agricultural works. This is a picture of farming *as it should be*, practised according to the principles of the new husbandry.

farming, for the most part at the end of the period. Of the numerous publications dealing with pure agriculture it was necessary to select for analysis only the most valuable ones. Nor has it been possible to include references to a number of studies published since the completion of this work, though some of these are listed in the bibliography.

It is a pleasant duty to express my gratitude to scholars who guided my steps when, at the end of the war, I had the unique privilege of resuming my work under their direction in the inspiring atmosphere of Cambridge. I was fortunate enough to have my research supervised by Professor Herbert Butterfield whose kindness to his disciples is too well known to be recorded. Professor M. M. Postan's help in the later stages of my work was an invaluable asset and it was his encouragement that made this publication possible. I am grateful to the Syndics of the University Press for their willingness to include it in their series of Cambridge Studies in Economic History. Finally such degree of readability as it has achieved is due to its revision by my two Cambridge friends, Miss Ursula Somervell and Mr Anthony Snellgrove. For their patience and devotion I cannot thank them enough.

May 1951   A.J.B.

PART ONE
THE PREPARATORY PERIOD
1700-50

CHAPTER I

FRENCH AGRICULTURAL LITERATURE
BEFORE 1750

IT IS quite remarkable that the important groups of studies on French agriculture in the eighteenth century lay the emphasis on agrarian theory; on the political, and even philosophical, aspect of the question, rather than on the technical. The authors of the standard works on these subjects are chiefly concerned with questions like the repartition of land, its mode of tenure, agricultural taxes, consequences of the feudal regime, and the connection of agriculture with the broader problems debated in that century, relating to political economy or statistics.

This tendency in the French agricultural literature of that period, though curious, is understandable. Questions of rural economy soon tended to leave the field of political economists for that of a more restricted group of specialized agricultural scientists. While both approaches to the question co-existed in the eighteenth century, the history of Agriculture in this period has often been treated by giving this word the wide meaning it possesses in English, whereas the history of Farming (a word for which there is no equivalent in French) has too often been disregarded.

In the eighteenth century itself, no distinction was drawn between them, since both studies were in their infancy. The public, not discriminating between the two tendencies, praised writings on both subjects with equal appreciation. The writings of the French agriculturists, however, have attracted the historians less than the more brilliant literature of the economists, with whom they have often been confused. This attraction is intelligible, in that the

3

tradition of economic history writing in France derived from the physiocratic movement, whereas the tradition of the agricultural literature was continued in more purely scientific works. The former led to the interpretation of the agrarian features of eighteenth-century France and their general economic consequences; the latter, to the elaboration of agriculture as an applied science. To their contemporary readers, both points of view were of equal interest; but discrimination between them today affords us a most interesting angle from which to study the period 1750-89—a period of important discoveries and of intense experiments, anticipating certain features of French agriculture of the following century.

The whole question seems, in fact, to have been sometimes misunderstood. In drawing one's information from the statements of the economists, the resultant picture of the state of French agriculture at this period is rather gloomy. Half of the kingdom is shown to be in a very poor state, the soil barren and the production of corn enormously decreased since the end of the seventeenth century. Moreover, the most up-to-date studies, based on statistics, tend to accept the view of an economic collapse before 1789.[1] Other views, however, may be found in other contemporary sources.[2] A certain author, famous in his day, thus celebrates the revolution of agriculture in that same century:

What a great number of important things there are to be observed and remembered! Not to mention many a minor item, all interesting and important, there is the history of grinding, of mulberry trees and of silkworms, of dye-plants, of potatoes above all ... not to speak of the history of veterinary schools, or horse-breeding and of studs, of the economic and agricultural history of wood growing. . . .[3]

[1] On these problems see G. Weulersse, *Le mouvement physiocratique en France*, Paris, 1910, and Labrousse, *La crise de l'Economie française à la fin de l'Ancien Régime et au début de la Révolution*, Paris, 1944.

[2] Article: Agriculture, *Dictionnaire des Sciences Naturelles . . . par plusieurs Professeurs du Jardin du Roi*, Paris, 1816, vol. I, p. 320. Article: Agriculture, *Nouveau Cours d'Agriculture*, Paris, 1809, vol. I, p. 168.

[3] *Cours d'Agriculture . . . par M. le baron de Morogues*, Paris, 1840, vol. I, p. 88, note 4.

4

The disagreement is merely superficial and not deep-rooted. We shall be helped to understand it by considering the fact that two different words were used to designate those engaged in the study of so complex a thing as agriculture. This is important, showing that they were not all interested in the same aspect of the problem.[1] The agronomes were concerned mostly with farming and to a much lesser extent with general economics.

This distinction explains the two different views expressed on French Agriculture.

It is certain that the form of government as it existed in France, together with its social system, its system of taxation, of economic regulations, the continual intervention of the State in the economy of society and the innumerable restrictions imposed on it, were the direct causes of the neglect and sluggishness in cultivating the soil[2] and of the peasant's indifferent exploitation of his land; in a word, of the rather poor returns of agriculture, when considered as one of the mainstays of the kingdom's economic structure.

It is also true to say, however, with Sagnac, that at the end of the *ancien régime*, agriculture, 'encouraged by better methods', was making forward strides.[3]

---

[1] This distinction in terms has drawn the attention of F. Brunot, *Histoire de la Langue française des origines à 1900*, Paris, 1930, vol. VI, p. 196. M. Brunot notes 'Agromane' in 1771 only. Fréron's *Année Littéraire* (1761) reviewing *L'Agronome ou Dictionnaire portatif du Cultivateur*, Paris, 1760, remarks, 'L'auteur donne à l'agriculture ou à l'administrateur d'un bien de campagne, le titre d'Agronome comme on appelle Astronome celui qui observe les astres.' Butel-Dumont, in *Recherches historiques sur l'administration des terres*, Paris, 1779, writes more precisely, 'Le mot Agronome a été introduit depuis quelques années dans la langue française pour signifier celui qui enseigne l'Agriculture, ou qui traite de ses règles, ou même seulement qui les a bien étudiées et qui en possède la science'. The word has therefore a more scientific implication which explains why it survived longer than 'Physiocrate' in French. It became extremely popular. The confusion has, however, been often made in modern writings.

[2] See innumerable statements in Weulersse, *op. cit.* vol. I, pp. 317 ff., 451 ff.

[3] Sagnac, 'La fin de l'Ancien Régime et la Révolution Americaine', *Peuples et civilisations*, Paris 1941, vol. XII, pp. 432, 433.

The teaching of the agronomes, although running parallel to that of the physiocrats, was nevertheless definitely distinct from it. More than the latter, the agronomes were experimental scientists,[1] and one may see, just as in other sciences such as botany, chemistry and physics (all of which contributed to their developments), a very real progress in the agricultural methods and technique between 1750 and 1789; or rather, a considerable advance in agriculture from the technical point of view.[2]

Even before 1750 the common root is visible which, branching out into science and political economy, ultimately flowered into a separate agricultural school. This common origin may be found in the great intellectual movement at the dawn of the eighteenth century, which was strongly marked with an English influence.[3] English thought penetrated to the French learned circles, and one of its most important forms was Political Economy. Ten years before Voltaire revealed England to the French, the famous 'Club de l'Entresol' was studying questions embracing economics and politics, under the inspiration of the English 'papers' and of aristocratic English lecturers whose influence was great.[4] It has been shown how agriculture came to be one of the main objects of the movement.[5] After 1750 it possessed a life of its own, but it is

[1] F. Brunot, *op. cit.* p. 216. Duhamel continually speaks of 'champ d'expérience'; Sarcey de Sutières writes an *Agriculture expérimentale*, Paris 1765. Home claims that agriculture can be reduced to a 'système régulier', *Les Principes de l'Agriculture et de la Végétation*, Paris, 1761, p. 7, also p. 262.

[2] Musset-Pathay writes about the year 1750: 'Alors l'agriculture tint un rang parmi les sciences, et comme elles, eut sa méthode et un corps de doctrines', in *Bibliographie agronomique*, 1810, p. 9.

[3] Paul Hazard, *La Crise de la Conscience Européenne (1680-1715)*, Paris, 1935. On relations between the different sciences in the eighteenth century, see G. Cuvier, *Histoire des Sciences naturelles*, Paris, 1843, vol. IV, pp. 185 ff.

[4] E. R. Briggs, *The Political Academies of France in the Early 18th Century, with special reference to the 'Club de l'Entresol' and to its Founder, the Abbé Alary*, Cambridge, Ph.D. dissertation.

[5] Weulersse, *op. cit.* vol. I, pp. 27 ff., also Weulersse, 'Le mouvement préphysiocratique en France', *R.H.E.S.*, Paris, 1931, pp. 244 ff. in which the common English source of the physiocrats and the agronomes is clearly shown.

well to remember that the English origin of the agricultural movement can be traced back to the beginning of the eighteenth
century.

Nevertheless, it would not be wholly correct to associate French
agronomy solely with experimental science and physiocracy, both
of which were related to English thought. This agriculturist movement was afoot at the beginning of the century, before Locke's
and Cantillon's works and the 'Club de l'Entresol' became fashionable in France, for a French tradition in agricultural writing was
in existence. The number of works on husbandry published in
France up to 1750 was relatively insignificant. There are a number
of authors, however, whose names have been recorded in the
annals of French agriculture, not only as representatives of á
certain kind of writing, but also as symbols of a certain French
craftsmanship. By the beginning of the eighteenth century, the
time had passed when the King indulged in listening to the reading of speeches in favour of agriculture[1] which were at the same
time valuable technical works. (Some instances are Olivier de
Serres' *Théâtre d'Agriculture* or Charles Estienne's *Maison Rustique*,
which have gained lasting fame for the advice they contain.)
Louis XIV's reign, in spite of Colbert's efforts in rural matters,[2]
was almost wholly engrossed in financial and trading problems.
Agriculture throughout this long period produced only mediocre
results, except in the annexed provinces, such as Flanders and
Alsace, where technique was advanced. In fact, at the beginning
of the eighteenth century, even though all the French provinces
may not have been as barren as La Bruyère depicts them, French
agriculture was indubitably at a standstill.[3] The efforts of Henry IV
and Sully failed, and only a century later, widespread indifference

---

[1] In G. Fagniez, *l'Economie sociale de la France sous Henry IV, 1589-1610*,
Paris, p. 39.
[2] E. G. Lodge, *Sully, Colbert and Turgot*, London, 1931, p. 162.
[3] A picture of this rather tragic situation can be found in Dom Leclerc, *Histoire
de la Régence*, Paris, 1922, vol. i, pp. lviii ff.

set in, which however, did not go as deep as it appeared to. Certain circles still retained an interest in agriculture. The gap between the last famous agricultural treatise (Liger's *Nouvelle Maison Rustique*, 1702) and the agricultural revival of the seventeen-fifties, is not as wide as has been made out.[1] A number of such technical books are to be found listed in contemporary bibliographies;[2] and although the agricultural writers were naturally less numerous in France than they were in England at this time,[3] the fact remains that as attention slowly shifted from pure politics to economics (as with Fénelon, Vauban, Boisguillebert, Boulainvilliers, Argenson, etc., all of whom were concerned with agriculture) interest in the technique of farming leant more and more for support on this early movement, until the discovery of English husbandry in 1750. Besides, even though no major work on husbandry appeared between 1702 and 1750, an unbroken movement, no matter how limited, in agricultural research runs throughout that period.

Since Colbert's *ordonnance* on the *Eaux et Forêts*,[4] a staff of specialists, often very competent men, was set up, from whom many an agronome of that time derived his knowledge of agricultural questions.[5] The King's taste for gardens brought into being a school of gardeners; disciples not only of Le Nôtre, but also of La Quintinie were working throughout France. The latter planned

[1] This is the opinion of Wolters, *Agrarzustande und Agrarprobleme in Frankreich von 1700 bis 1790*, Leipzig, 1905.

[2] The Abbé de Petity, in his *Encyclopédie élémentaire* (3 vols), Paris, 1767, gives not less than 10 titles of books on agriculture, published between 1700 and 1750 in France (vol. 1, pp. 582 ff.). Many other titles will be found in the *Catalogue* of Huzard's library.

[3] See G. E. Fussell, *The Old English Farming Books*, London, 1947; also Paul H. Johnstone, 'In praise of husbandry,' *Agricultural History*, vol. XI, no. 2, 1937, p. 85. [4] See below, p. 122, n. 3.

[5] For instance, Duhamel, Le Roy (who writes in the *Encylcopédie*), also many a minor one, like Thierriat, 'Conseiller du Roy, Garde Marteau de la Maîtrise des Eaux et Forêts de Chantilly', who writes in 1764 *Instructions familières . . . sur la Culture des Terres*.

a kitchen-garden at the King's command—the best example of its kind in Europe. This type of garden was much cultivated at first, and its products exported all over Europe.[1] The knowledge of these scientific gardeners was not confined merely to raising crops, but included general problems of agricultural and vegetable physiology; they were proud of their art, the dignity of which they praised:

Agriculture in general, may be look'd upon as a Science of vast extent and proper to afford Philosophical wits an infinite deal of Exercise, no part of Natural Philosophy yielding more excellent matter for contemplation, or being more fertile in useful and delightful experiments than that which treats of Vegetation. For I known there are abundance of fine and curious questions proposed in it; as for instance whether the sap circulates in Plants as the blood does in animals? Whether the roots do actively attract or only passively, without any action on their side, receive the juice which serves for the nourishment of every Plant. . . .[2]

Some time later, the Abbé de Vallemont wished that 'this important knowledge might spread among the people in the country' and that he might 'cause everything useful which had long since been discovered, to pass from Scientists unto the People'.[3] This shows an interestingly fresh point of view, quite different from that of the more archaic authors, like Liebaut or Serres, and heralding that of the agronomes.

Besides these Royal or private gardens, which kept alive an uninterrupted tradition of research and experiment in France, there was a most important centre—the 'Jardin du Roi'—which

[1] The famous orchards and kitchen-gardens of Montreuil near Paris are linked with La Quintinie's efforts. De Pradt, *De l'état de la Culture en France*, Paris, 1802, vol. I, p. 30, insists on the importance of this fruit growing, initiated at the end of the eighteenth century. No doubt the pupils of La Quintinie must have had some influence when working in the provinces for the nobility.

[2] La Quintinie's *The Compleat Gard'ner*. Translation by John Evelyn, London, 1693, Preface.

[3] Vallemont, *Curiosités de la Nature et de l'Art sur la Végétation, l'Agriculture et le Jardinage dans leur perfection*, Paris, 1703.

was to prove extremely serviceable in agricultural research, the value of which was increasingly appreciated throughout the eighteenth century, while its links with the revival of husbandry were numerous. Although the 'Jardin du Roi'[1] was at first in the care of pure scientists, like the botanists Tournefort or Du Fay, it counted among its *pensionnaires* people like Duhamel du Monceau whose work on agriculture was to prove so important. In this garden, the anatomy and physiology of plants were methodically studied, species were classified, compared and cultivated. This research did much towards improving agricultural methods later.[2] All the scientists who were then working in the 'Jardin du Roi' were either members or associate members of the Royal Academy of Sciences. For during the greater part of the eighteenth century, husbandry was studied only by the Academic Boards of Chemistry, Botany or Zoology; it was not yet recognized as a science in itself.[3]

It was also thanks to these two centres—the 'Jardin du Roi' and the Royal Academy of Sciences—that the English influence, the extent of which will be analysed in the following chapters, penetrated into the French scientific world interested in agriculture. The great controversy between Cartesianism and Newtonianism, had acquainted the majority of French scientific circles with English trends of thought. Acquaintance with English botanists, zoologists and chemists[4] preceded the discovery of English

[1] Later the 'Jardin des Plantes', see *Annales du Muséum*, 1804. Historical notice by A. L. de Jussieu, vol. III, pp. 3-17; vol. IV (1805), pp. 4-19.

[2] 'Botany and Agriculture are a mutual support to one another; the one is the principle of the other; the latter works in order to prove the former useful.' *Contemporary Treatise*, quoted by Calonne, *La vie agricole sous l'Ancien Régime en Picardie et en Artois*, Paris, 1883, p. 48.

[3] It was constituted as such only after the example of English thought, mostly that of Bacon. See 'Observations sur la division des Sciences par le chevalier Bacon', *Encyclopédie*, 1751, vol. I, pp. li-lii.

[4] See important bibliography for the eighteenth century in Ascoli, *La Grande Bretagne devant l'opinion française au XVIIe siècle*, Paris, 1930, t. II, pp. 297 ff. An instance of this exchange of scientific knowledge may be found in Robert

agriculture. Du Fay, at the beginning of the eighteenth century, reorganized the 'Jardin du Roi', after a journey to Leyden and to England, chiefly, where he visited the Hampton Court, Chelsea, and Eltham Botanical Gardens;[1] the masters of the young Duhamel were the brothers de Jussieu whose works were written on the lines of those of their English colleagues; Bernard de Jussieu came to England in 1727 to purchase plants;[2] the new school of French zoologists, which supplemented the studies undertaken by John Ray (until 1734 the only well-known zoologist in France)[3] was in various ways connected with that of England; numerous articles by people like Hans Sloane were read at the Academy and discussed by Réaumur or Hérissant or Brisson; the Academy sent its younger members on official visits to England—Duhamel, to study ship-building; Jars, later, to study metallurgical processes.[4] All this preparatory work was brilliantly epitomized in 1749 in Buffon's *Natural History*; and Daubenton's unmentioned collaboration in the work (Daubenton, whose research work on sheep breeding later played a leading part in the new agricultural science) is another link in the chain binding the French agricultural movement to the scientific school briefly introduced above.

Outside the scientific world, interest in agricultural problems is met with in isolated cases. The works of the Marquis d'Argenson

Morison, F.R.S., born in 1620, who, after his studies in Paris, 'was recommended by Mr Robins, the French King's botanist, to the patronage of the Duc d'Orléans, uncle to Louis XIV, who appointed him keeper of his fine garden at Blois with a handsome salary'. Thomas Thompson, *History of the Royal Society from its Institution to the End of the XVIIIth Century*, London, 1812, p. 28.

[1] Maury, *L'Ancienne Académie des Sciences*, Paris, 1864, p. 110.

[2] *Annales du Muséum*, 1804, p. 4. Fontenelle writes: 'La correspondance avec les étrangers qui fut le résultat de ces voyages, établit un commerce qui nous était d'abord désavantageux parce que nous étions ou dans la nécessité d'acheter, ou de recevoir des présents; mais on en vint par la suite à faire des échanges avec égalité et même enfin avec supériorité.' *Ibid.* p. 15.

[3] Maury, *op. cit.* p. 122. Also notice by Cuvier and Dupetit-Thouars in *Biographie Universelle* (Michaud).

[4] Cf. Ballot, 'La Fondation du Creusot', *Revue d'histoire des doctrines économiques et sociales*, Paris, 1912, p. 30.

or those of the Marquis de Mirabeau, although primarily concerned with social or economic problems, are yet closely related
to problems of practical agriculture and show a very real and
profound knowledge of them. Twenty years before Duhamel's
work on Quesnay's articles, we meet in d'Argenson sentences
which advocate the principles of the modernized agriculture as it
existed in England.

The rich citizen of a near-by town should not own a field in the
neighbouring countryside for what income it brings him, but in order
to improve it more and more, since in our vast and wretched provinces in the interior of the Kingdom, everything has been forced to
a standstill. Nothing has been brought up to date, the old methods
of cultivation are continued, and that slowly and with indolence.[1]

Again, he suggests that the country's profits should be sought in
a perpetual give and take between land and cattle, under man's
diligent supervision: 'Cattle graze on pastures, their manure growing new ones and making lands fertile. . . . Work and manure
make lands fruitful.'[2]

The Marquis de Mirabeau, in his famous *Ami des Hommes*,
expresses the same opinions. His book is not only a dissertation
on economics, it is also a practical treatise on husbandry, in which
he draws on experience acquired as a gentleman farmer, long
before the publication of his work in 1758. Another instance of a
study on the subject, devoid of 'philosophical' treatment, is the
Marquis de Turbilly's *Mémoire sur les Défrichements*.[3] The author
tells the agronomes how he applied new methods of farming in
his estate in Poitou. There are other lesser authors, whose work
should not be overlooked, and who are genuinely interesting.

[1] *Considérations sur le gouvernement ancien et présent de la France*, Amsterdam,
1764 (written 30 years before publication by the Marquis de Paulmy, his son),
p. 302.   [2] d'Argenson, *op. cit.* p. 301.
[3] Appeared in its English translation as, *A Discourse on the Cultivation of Waste
and Barren Lands*, 1762. A. Young and other English authors held it in high
esteem. First published in Paris in 1760. According to Turbilly (p. 151) the
improvements began in 1737.

The Abbé de Vallemont presented his book on agriculture to the Marquis de Dangeau, an amateur agriculturist, who belonged to the circle of which de Petity wrote: 'La culture des Terres labourable et des jardins, qui font aujourd'hui l'objet des soins et les délices des personnes curieuses et de la plus haute condition.'

The Maréchal de Noailles, a botanical convert, took up agricultural research on his lands at St Germain,[1] and was later to become a patron of Duhamel; the Duke of Orleans sponsored the translation of Tull's work in French; the banker, Paris de Monmartel, cultivated his estate at Brunoy as a 'great capitalist'[2] during the Regency period.

Briefly, there was in France immediately before 1750 both an agricultural movement and an agricultural literature, even though scattered. Experiments were carried out and agricultural problems gone into. At this time, the great agricultural revolution had begun in England, and its results were suddenly revealed in 1750. A gap between the years 1702 and 1750, as noted by Wolters,[3] there certainly was, but it was not as complete as has been thought so far. Agricultural research and writings had not altogether ceased during this period, although the old agricultural technique had been increasingly neglected[4] and the old 'Art' of Serres and La Quintinie had come to a standstill in the greater part of the

[1] Jussieu, 'Historique du Jardin du Roi', *Annales du Muséum*, 1805, vol. VI, p. 16 n. 1.

[2] Marc Bloch, 'Les Plans parcellaires', *Annales d'histoire économique et sociale*, Paris, 1929, p. 69.   [3] See above p. 8, n. 1.

[4] The old 'Rustick Authors' were only 'rediscovered' at the extreme end of the eighteenth century, being in great favour at the beginning of the nineteenth, when the agronomes began to escape the total influence of England. (See François de Neufchâteau's edition of *Olivier de Serres*, Paris, 1804.) The *Cours d'Agriculture . . .*, 1815, writes: 'car alors la France, tourmentée de l'anglomanie, laissait dans le plus profond oubli sur Olivier de Serres, et semblait ignorer que nos Flamands sont les maîtres de cette agriculture anglaise si vantée' (t. II, p. 512). In fact, the movement of the agronomes begins with the translation of English works. De Pradt, *De l'état de la culture en France*, Paris, 1802, vol. I, p. 7, complains in his Dedicatory Letter to Arthur Young that France has always lacked 'agricultural thinkers'.

kingdom. The new husbandry was to be discovered soon, for England, in contact with the different movements of the period, revealed her prosperous and modern husbandry to the French agriculturists. In 1750 Duhamel du Monceau was expounding the new methods discovered in England to the French.

CHAPTER II

THE DISCOVERY OF ENGLISH
AGRICULTURAL METHODS

IT WAS around the year 1750 that the French came to know the actual state of English agriculture. Until that time, the emphasis had been laid chiefly on political, social and 'philosophical' aspects of England. The manifestations of the country's life were known only to the purely intellectual or scientific circles. Gradually, however, following in Voltaire's footsteps, the number of travellers to England increased. They described and often published their impressions of a country still unfamiliar.[1] The general public was kept informed of England's political and social life, of its literary productions and of its scientific discoveries. In addition, various aspects of the English countryside were described (often, naturally, as a means of propaganda).

The Abbé Coyer, in his *Noblesse Commerçante*, writes that England had begun to be transformed 50 years before, although the results of the transformation had become apparent only 20 years earlier. In comparing the descriptions of French and English landscapes in the mid-eighteenth century, one may well imagine how keen an interest the latter must have awakened in their French readers. In contrast to their own country, where agricultural life was on the whole wretched, these new descriptions showed a picture of a veritable Garden of Eden, of a country 'always smiling'.[2] In certain parts of this country, the prosperous farming had made

[1] About this 'discovery' of England, see below p. 180, n. 1. Also D. Pasquet, 'La Découverte de l'Angleterre par les Français au 18e siècle', *Revue de Paris*, 1920, vol. VI; 1921, vol. I.

[2] In Abbé Le Blanc, *Letters on the English and French Nations*, London, 1747.

15

of them, less a *campagne* than an 'immense garden'. These descriptions corroborated Voltaire's almost incredible accounts of English peasants who 'eat white bread, are well cloath'd and are not afraid of increasing their stock of cattle, nor of tiling their houses from any apprehension that their taxes will be rais'd the year following'.[1] Thus England, the birthplace of philosophy and of the ideal government, could also boast of a wealthy and thriving countryside. Throughout that century and into the next one, England gained increasing admiration on that score.[2] In most of the books on economic or English questions, published around 1750, praise of English agriculture and farming is practically never absent.[3]

The archaic treatises on husbandry at the beginning of the century were lacking in details. Once or twice, however, in La Quintinie or Liger, examples of English methods can be found. In the interesting *Dictionnaire Oeconomique* of Chomel there are certain very precise hints of the new theories extracted from British authors. From several of his articles it appears that the author knew, and had read, most of the books which had been translated on botany or vegetable physiology. He quotes M. Digby and his *Traité de la Végétation de Plantes*[4] in support of his own theories. His article on the technique of seed sowing is inspired by Ray's precepts in his *Historia Plantarum*.[5] He was acquainted with English agricultural life and methods, through Camden's *Description de la Province de Cornouailles*. Chomel observes that in England 'there are farmers whose custom it is not

---

[1] Voltaire, *Letters Concerning the English Nation*, London, 1773, p. 68.

[2] Léonce de Lavergne, in *Economie rurale de la France*, Paris, 1866, is the typical representative of this state of mind.

[3] The great movement of transformation was only beginning and was not well known until some twenty years after. Anyway, the English landscape impressed its visitors then. About the impressions made on French visitors by the English countryside, see Ascoli, *op. cit.* vol. I, ch. III, part 5, pp. 330 ff.; D. Pasquet, *Revue de Paris*, 1921, vol. I, p. 835.

[4] Kenelm Digby, *A Discourse concerning the Vegetation of Plants*, 1669.

[5] London, 1686-1704.

to prepare wheat', while he himself writes much about the 'elixirs of multiplication'. He gives several methods of fertilization applied in Cornwall, and other counties[1] (which were later dealt with in the *Journal Oeconomique*), and explains how land is improved by burning the turf, or by scattering sea-sand and seaweed.[2] Chomel also makes some very sensible suggestions on the improvement of stock breeds in France and tells, for instance, how in England it was achieved by crossing English with Spanish sheep.

These still tentative reports on agricultural conditions in England were amplified as knowledge of the country grew more intimate. In 'the most delightful island of Europe', the orchards, the hedges enclosing the fields, even then made an immediate impression. The grass seemed to be more abundant and of better quality than anywhere else. More precise reports spoke of the quality of soils which had been improved with marling. Apart from a few hills in Wales, some sterile plains, such as that of Salisbury and some marshy counties, 'England is generally beautiful and fertile'.[3] The efforts of various individuals were held up as examples.[4] The intensive cultivation then beginning was noted by *L'Espion Chinois*: 'The whole of England is cultivated. Not a single part of the land is wasted. It is perhaps the only Kingdom in Europe which not only provides food to all its inhabitants, but

[1] In *Britannia: or a Chorographical Description of Great Britain and Ireland*, first published in Latin, then translated by Edmund Gibson in 1722.

[2] Already quoted by the eighteenth-century travellers, Coulon and Misson, who celebrate Cornwall 'd'un merveilleux rapport par le travail de ses habitants qui l'eng raissent de l'algue de la mer et d'un certain sable qui a les mêmes effets que le fumier'. [3] Misson, quoted by Ascoli, *op. cit.* p. 290.

[4] In Vallemont, *Curiosités de la Nature . . .* , *op. cit.* Preface, 'Mais du moins je souhaiterais qu'on prit plus d'intérêt à faire valoir ses terres, qu'on s'appliquât à perfectionner l'Agriculture et le Jardinage comme on a essayé de perfectionner les autres Arts qui sont bien moins utiles à la vie. Nous voyons dans les Actes Philosophiques de la Société Royale d'Angleterre, que les Grands, les Savants personnages qui la composent ont fait là dessus de belles découvertes'. Cf. G. E. Fussell, 'Husbandry in eighteenth-century England', *Agricultural History*, xv, 4, 1941, p. 213.

also supplies the means of life to other nations.'[1] One of the most popular writers of the period on these English themes is the Abbé Le Blanc. The Abbé Le Blanc, whose literary importance has been studied,[2] should also be considered for his contribution to agriculture. He is indeed the first to give accurate and attractive pictures of English farming and its results, with no ulterior economic motive.[3]

To Buffon, asking for a description of the country, he answers, praising a land 'as agreeable as it is fertile', that, 'Whoever has eyes, must be struck with the beauties of the country, the care taken to improve lands, the richness of the pastures, the numerous flocks that cover them, and the air of plenty and cleanliness which reigns in the smallest villages'.[4] The reason for this rich and pleasant aspect of the country? It is because, says the Abbé, the English do not concentrate in London, their capital, which is devoid of attractions, but they inhabit the country, where rich landlords can look after the cultivation of their estates. ''Tis only in the country that they display all their magnificence . . . the rest of their leisure, some spend in applying themselves to agriculture, and the pleasures of domestic life.'[5]

Le Blanc, who is not associated with the economists, yet sets forth the same arguments as theirs, merely as a result of his English experiences.

What makes the English love planting more than we do, is that those who by birth or riches, are of the greatest distinction in the State, live

---

[1] *L'Espion Chinois*, Cologne, 1744, vol. IV, p. 5. (Anonymous. Perhaps by Ange Goudar, political satirist, also author of *Les Intérêts de la France mal entendus dans les branches de l'Agriculture*, 1756.)

[2] H. Monod Cassidy, *Un Voyageur philosophe au XVIIIe siècle, J. B. Le Blanc*, Cambridge (Mass.), 1941.

[3] *Lettres d'un Français sur les Anglais*, 1745. Several other editions, 1749, 1751, 1758.

[4] Le Blanc, *Letters on the English and French Nations*, translated from the original French, London, 1747, vol. I, p. 177.

[5] *Ibid.* vol. I, pp. 262-3; The French text is 'et aux charmes de la Vie Economique'.

in the country more than those of the same rank in France. . . . As the nobility set the fashion to their inferiors, so the farmer plants in imitation of his landlord. . . . In our villages, the farmers seldom plant anything but apple-trees and cabbages; the English farmer has not only a kitchen-garden well furnish'd and kept in good order, but if there are two fathoms [sic] of ground before his house which belong to him, he makes a flower garden of it, where he cultivates the rose and lily of the valley; a sufficient proof of his easy circumstances. They seldom employ themselves in cultivating flowers, but when there seems to be a promising appearance of a good harvest.[1]

He then holds up the English farmer as an example to the French landlords and points out the material benefits accruing to the former.

'Agriculture is a slow but sure way of enriching oneself; the earth rewards those who cultivate her with her produce. The English, in this respect, more sensible than we are, look upon this as the principal means of augmenting their estates. . . . How many men would not enlarge their fortunes if they follow'd the wise example the English set us?'[2] Nor is this state of thing derived from purely economic or political reasons. English literature on farming and husbandry is to Le Blanc's eyes as valuable as it is abundant. 'As among the Romans, a Cato did not disdain to write upon agriculture, so we must own to the honour of the English that some of their most eminent authors have published very instructive works on this subject. . . . No other nation has so many useful books on this subject.'[3] The Abbé shows further the important part played by the Royal Society of London, implying that the Royal Academy of Sciences of Paris must allot itself the same role. 'We must acknowledge to the honour of the Royal Society of London, that 'tis its continual attention to the public good, has procur'd England all these advantages.'[1] And the Abbé sees, not without grounds, in Buffon, a typical representative in France of the English methods. 'It will not be your

[1] *Ibid.* vol. I, p. 321.   [2] *Ibid.* vol. II, pp. 59, 60.   [3] *Ibid.* vol. I, pp. 318-19.
[4] *Ibid.* vol. I, p. 322. Also see D. Pasquet, *Revue de Paris*, 1921, vol. I, p. 214.

fault, Sir, if we don't follow the wise example of our neighbours.'[1] And again, 'You certainly set them an example worthy of imitation when purely out of love to natural philosophy, and to facilitate the progress of those who study it, you were pleased to interrupt your particular occupations, to translate Dr Hales's Vegetable Statics, the best author the English have on this subject.'[2]

The material results Le Blanc presents are interesting in that they show English rural life as seen through French eyes, and are at the same time a lesson to his agronome compatriots.[3] He chooses cleverly what will strike his reader. To Quesnay, the symbol of a rich agriculture was ploughing with horses; to the Abbé Le Blanc and his readers it was the farm servant having tea before going to the fields.[4] 'The care with which the country is cultivated there is the consequence of the plenty in which the farmer lives. In England as well as in Holland, the villages are neater and better built than in France. The effects of this wise economy are visible in everything in the country, even in their animals, and the earth repays the husbandman with usury what it costs him to have good horses and feed them well.'[5] The quality of the crops and, in general, the quality of many agricultural

---

[1] Le Blanc, *Letters on the English and French Nations*, vol. I, p. 323.

[2] *Ibid.* vol. I, p. 319; A. L. de Jussieu, in *Annales du Muséum*, 1805, p. 2, also ascribes to Buffon a translation of Newton, and of Tull's *Essays on Husbandry*.

[3] These pleasant descriptions of the English countryside were written at a time when the agrarian destinies of Great Britain were on the verge of changing. It is not probable that the Abbé Le Blanc refers, in his praise of English agriculture, to the very specialized change that was taking place then in Norfolk. It is more likely that the richer or wealthier aspect of the English landscape called for his attention; and this feeling will be generally of great importance to accredit English methods in France, even outside any political or social explanations.

[4] The Abbé Coyer, in *Développement et Défense du système de la Noblesse Commerçante*, Amsterdam, 1757, also points out this luxury in the life of peasants. In England, 'Le laboureur briserait sa charrue s'il était obligé de vivre comme le nôtre, de noix, de châtaignes, d'orge ou de blé noir', p. 32, 2nd part. Young refers frequently to the habit of tea-drinking among agricultural labourers, *Tour through the Southern Counties*, p. 210; *Tour through the North*, vol. III, p. 339; vol. IV, p. 351.   [5] Le Blanc, *op. cit.* vol. I, pp. 297 and 298.

products, testifies to the care with which the English cultivate their country, and to their improved methods of cultivation. The Abbé then enumerates some of these products.

Sir, I shall have the pleasure to satisfy you, and entertain you today with the taste which the English have for gardening and plantations; and of the wonderful effects this taste has produc'd in their country. I shall say nothing of their ability in this respect . . . and indeed fruit and pulse are nowhere cultivated with so much care and industry. Though the climate is not so favourable here, as in France, they have carried the art much farther. You find in the markets at London green peas more early than at Paris, and pineapples at all seasons—and various sorts of pulse, which we have not, are very common. Broccoli which is still so rare with us, they eat here in the publick houses. . . . You do not only find fruit trees of all countries in England, but you also find a prodigious quantity of those trees, which have no other worth but their beauty, or the singularity of their form. The English import all sorts of trees, at a great expense, from different parts of the world, and those which thrive in the open air, they naturalise.[1]

This, according to the Abbé, is the result of making agriculture an object of care and study. The admiration he shows for the English writers of books on husbandry is very characteristic: 'We see by all these books on gardening that it must be better understood here, than anywhere else.'[2] The science of agriculture shows wonderful results in a country where a large part of learned opinion is interested in such matters. This is the typical attitude of the agronomes.

The Abbé Le Blanc is but one example of those pioneers of Franco-English agricultural relations. Such remarks may be found in other works of the same school—the Abbé Prévost's, for instance.[3] But Le Blanc's are particularly suggestive.

Some time later, in 1751, in the most popular work of the

---

[1] *Ibid.* vol. I, pp. 316 ff.
[2] *Ibid.* vol. I, p. 319. Same remark in Patullo, *De la Clôture des Terres, Essai*, p. 3.   [3] *Mémoires et aventures d'un homme de qualité*, 1757.

eighteenth century,[1] Diderot's *Encyclopédie*, English agriculture makes another appearance. Eleven years after the 1740 edition of N. Chomel's *Dictionnaire Oeconomique*, where it is so briefly mentioned, it takes the first place in the economic articles of the *Encyclopédie*.[2]

Diderot's second agricultural article, especially, 'Culture des Terres' published in the fourth volume three years later, he devotes entirely (if we except his purely economic reflections on the Corn Trade, in which he strongly advises imitation of England) to a picture of English agriculture through a *Lettre écrite du Comté de Norfolk*.

It must be pointed out that in those days it was Norfolk which was the best known and most described of the English counties. This fact is doubtless due to the personality of the Norfolk pioneer, Lord Townshend, a character popularized by Voltaire[3] and the type of the 'grand seigneur agronome'. Around 1750 it is interesting to note that these Norfolk Letters become more and more numerous. Besides that of the *Encyclopédie* there is one in Veron de Forbonnais' *Eléments du Commerce*[4] and many extracts from it in the various periodicals of the time.[5] The French authors are particularly interested in the case of Norfolk, in the fact that a region of sterile soil and poor husbandry was transformed thanks to the bold undertakings of a great landlord using new agricultural methods.[6] Diderot speaks of this letter as follows: 'instructive

---

[1] D. Mornet, *Les origines intellectuelles de la Révolution Française*, Paris, 1933, pp. 75 ff.

[2] Wolters, *op. cit.* p. 169, does not sufficiently show the difference between the tone of the two articles ('Agriculture' 1751 and 'Culture des Terres' 1754). In the former there is some scepticism about Duhamel's conclusions. In the latter, it is the triumph of the 'new husbandry'.

[3] And St Lambert in *Les Saisons*, vol. III, pp. 142-4.

[4] 'Etat de l'Agriculture dans le Comté de Norfolk et de la méthode qu'on y suit', *Eléments du Commerce*, Paris, 1754, pp. 128 ff.

[5] *Journal Oeconomique*, Jan. 1755, pp. 145-6; *Nouvelliste Oeconomique et Littéraire*, July 1756, vol. I, p. 134.

[6] Lord Ernle, *English Farming*, p. 173.

pour les terres de même nature qui peuvent se rencontrer parmi nous', and the *Nouvelliste Oeconomique* says, 'La lettre Anglaise dont on donne la traduction sur les améliorations faites dans le comté de Norfolk est frappante et très propre à exciter l'émulation parmi ceux qui possèdent des terres en France'.[1]

The remarks on agriculture in Norfolk are of course more precise than those of Le Blanc, and some of the main concerns of the agronomes appear in them.

First, questions on the nature of soils and fertilizers needed, as, for instance, the mixing of two soils of different kinds. 'On croit . . . que feu Milord Townshend a le premier imaginé de féconder nos terres avec de la glaise.'[2] Certain manures, unknown in the *Maison Rustique*,[3] are complacently quoted because of their obvious strangeness: 'L'engrais de ces terres est le parcage des moutons, le fumier ordinaire, de vieux chiffons, des rognures de drap qu'on coupe en très petits morceaux et qu'on jette sur la terre immédiatement après qu'on a semé.'[4] They were to discover that the English used sea-water itself, salt and shells, as fertilizers. French writers point out that it is possible to 'make heath lands fertile' when deeply ploughed and sown with 'turnipe'. They quote recent improvements in Hanover, brought about with this seed. They conclude: 'Ce que le roi d'Angleterre propose à ses sujets, nous le proposons à nos concitoyens.'[5]

The question of new plants is also one of growing concern. Not only are the qualities of turnips discovered; but also those of a great number of other plants either entirely new or long forgotten. The order in which these new crops follow each other is also

[1] At the end of the century Baert will still particularly mention Norfolk. A. Young, however, claimed to have revealed Norfolk husbandry to his contemporaries. In *Autobiography*. Ed. M. Betham-Edwards, p. 44.

[2] *Jour. Oecon.* 1751.

[3] The *Maison Rustique* is the standard agronomic work of the eighteenth century, the book to which one alludes and from which one borrows constantly.

[4] 'Culture des Terres', *Encyclopédie*, vol. IV, 1754, p. 563.

[5] *Jour. Oecon.* 1751, vol. I, p. 70.

itself a novelty. So is the close connection in England between soil-cultivation and livestock, and resulting from it the whole question, that was later to acquire such importance, of artificial grasses, one of the 'grandes richesses de l'Agriculture Anglaise'. This sort of agriculture,

ne sépare jamais la nourriture des bestiaux du labourage, soit à cause du profit qu'elle donne par elle même, soit parce qu'elle même fertilise les terres: ainsi alternativement une partie des terres à blé d'une ferme est labourée en grande et petite luzerne, en trèfle, en sainfoins, en gros navet dont il paraît que nous conservons le mot anglais turnip . . . enfin, une herbe qu'ils appellent ray-grass, qui est inconnue à nos cultivateurs puisque nous n'avons pas de mot pour la rendre.[1]

Instances of these queries about new plants might be easily multiplied. So too with livestock. Agriculture and cattle breeding being intimately connected, it is no wonder that praise was applied to cattle as well as to the land. French authors were struck by the abundance of livestock, especially of smaller beasts: 'In order to give you some idea of the surprising multitude of flocks of sheep raised in Great Britain, be it sufficient to say that the annual number of fleeces cut is more than 30 millions.'[2] The methods of breeding such a large number of animals also began to awaken interest. Oxen from Buckinghamshire, Lancashire, and Somerset were said to be particularly remarkable.[3] This was due to the care with which they were treated when young: 'The English feed their calves with much attention; the care they take of them is almost an art in this country.'[4] Later we shall find such remarks in Daubenton.

It is, then, the appearance of wealth and prosperity in English agriculture which awakens the interest of a certain circle of French

[1] 'Culture des Terres', *Encyclopédie*, vol. IV, 1754, p. 563.
[2] *Année Littéraire*, 1755, vol. II, p. 73.
[3] M. Dumont, *Essai sur le Commerce d'Angleterre*, Paris, 1755 (translated from J. Cary and enlarged by the translator) p. 74.
[4] *Ibid.* p. 74.

writers interested in rural questions. These, however, study only the technical aspect of the question. Others, the physiocrats, will attempt to explain this technical superiority, to understand its causes. Within our purely agronomical school, interest was differently focused. Its first authors were longing for a better cultivation of the soil; and now rich landscapes and ingenious methods were revealed. A few years after 1750, England as a school of agriculture was an accomplished fact; and in order to acquire a more intimate knowledge of the agricultural world of the country, a vast enquiry began. Economic periodicals and papers began to publish extracts from English papers dealing with those questions. They were evidently a great success, for the Editor, in a note, thanks the public: 'Nous apprenons que le Public a reçu avec satisfaction les Extraits des Journaux d'Angleterre que nous lui avons présentés jusqu'à présent et la promesse que nous lui avons faite en même temps de continuer toujours à donner ainsi chaque mois ce qu'il y aura dans ces sortes d'ouvrages anglais de capable de lui plaire et de relatif au genre auquel nous nous sommes voués.'[1] He seems also to have roused some protests (which nevertheless give the English nation full credit for its qualities), for an anonymous writer, in a letter to the Editor, finds that 'one must be very careful in choosing among English books; each fashion of thinking has its own time; Anglomania may now be reigning. . . .'[2]

In comparison, however, with the increasing influence of England, these reactions (which are the symptom of a particular tendency)[3] in no way diminished the respect surrounding the names of the great English scientists. Names, which 20 years before, were known only to a restricted circle of intellectuals

[1] *Jour. Oecon.* 1751. About these 'English Papers' see Fréron, *Année Littéraire*, 1750. A list of 24 of those obtainable is given.

[2] *Jour. Oecon.* 1753.

[3] There is undoubtedly a tint of anglophobia in some of the reactions to the new influence (see below, ch. v, p. 61, n. 4).

and Academicians, gradually became household words. Books were full of such epithets as 'celebrated', 'learned', 'esteemed', addressed to M. Philip Miller or M. Ellis. Dr Hales was always 'worthy'.

All this indicates the tendency to admire English achievement in the field of agriculture. It also shows the beginning of a better acquaintance with their methods. The story of Jethro Tull's agricultural treatise and its French destinies is a picturesque example of the interest shown by some members, and those not the least, of French Society in such English works.

On en [Tull's book] parla bientôt en France. M. le Maréchal de Noailles, toujours attentif au bien public, engagea feu M. Otter de l'Académie des Belles Lettres, à traduire en notre langue le traité de M. Tull. Pour rendre parfaitement le sens de cet ouvrage, il ne suffisait pas de savoir parfaitement l'Anglais, il fallait encore aussi avoir quelque connaissance de l'Angleterre. C'est pourquoi M. Otter ayant fini sa traduction, pria M. de Buffon de l'examiner. Celui-ci se chargea de cette révision qui l'occupa quelques mois. Pendant ce temps on envoya à M. Duhamel une autre version du même ouvrage faite par M. de Gottford. Elle était accompagnée d'une lettre de feu M. Le Chancelier qui le priait de faire sur cette seconde traduction la même révision que M. de Buffon avait faite sur celle de M. Otter. Les deux Académiciens se communiquèrent leurs idées sur le livre de M. Tull. . . . M. Duhamel jugea que pour mettre les amateurs de l'agriculture à portée de profiter de celle-ci, il fallait se contenter d'en faire un bon extrait. . . . Chargé de ce travail, M. Duhamel entra dans les vues de M. Tull, y mit l'ordre qui lui parut convenable . . . apporta des expériences pour confirmer le sentiment de l'Auteur.[1]

This work which excited very wide interest is actually the first methodical exposition of English (or of *an* English) agricultural technique.[2] Hitherto the descriptions, however enthusiastic, had been vague and general. From now on, agriculture was to

---

[1] *Nouvelliste Oeconomique et Littéraire*, 1755, vol. IV, p. 4. *Journal des Savants*, Oct. 1750, pp. 330 ff.

[2] M. Bloch, 'La vie rurale jadis et naguère', *A.H.E.S.* 1930, p. 102.

constitute an experimental science,[1] and the English were to show the French agronomes how to set about improving their rural system. This was, in general, Duhamel's object in translating and adapting Tull's work.

---

[1] M. Bloch, *Les Caractères originaux de l'Histoire rurale française*, p. 222, is very right when he includes the 'agricultural revolution' within the movement of thought in the Century of Enlightenment: rationalization of practice, contempt for tradition. However, it seems that the 'new husbandry', as it was taught by the agronomes, may be included mostly in the movement of experimental science which owed so much to England.

PART TWO
DUHAMEL DU MONCEAU'S WORK

CHAPTER III

THE OLD FRENCH HUSBANDRY

BEFORE giving an account of the Tullian system as it was to be interpreted by Duhamel, it seems necessary to say a word on the old French agricultural system before 1750, and the 'ancienne manière'. This old rural economy based on the open-field system has been well pictured by Lord Ernle[1] and, on the whole, his description can be applied to France. But certain technical and social conditions can be met with in French agricultural writings of the pre-Tullian period which give an original position to the French agrarian problems. Modern historians, too, have indeed given excellent accounts of it,[2] and we shall therefore picture this situation in broad terms only; attempting nevertheless to set in full light the defects of an organization against which the agronomes were later to struggle. However limited, the picture seems a necessary background to the work of Duhamel and to its full understanding.

The old French agricultural system rested on two foundations. First, the primary importance of the cultivation of cereals[3]—

---

[1] *English farming past and present*, pp. 154 ff. See also J. L. and Barbara Hammond, *The Village Labourer, 1760-1832*, London, 1919, pp. 25 ff.

[2] Marc Bloch, *A.H.E.S.* p. 334; H. Sée, *La vie économique et les classes sociales en France au XVIIIe siècle*, Paris, 1924; Octave Festy, *L'Agriculture française pendant la Révolution*, Paris, 1947 (Introduction); George Lizerand, *Le régime rural de l'ancienne France*, Paris, 1942.

[3] It may be pointed out in passing that whenever the old writers speak of 'bled' they do not always mean wheat, but also buckwheat or rye. This has sometimes led to inaccuracies in statements about the corn problem. For instance, in Bourgin, 'L'Agriculture et la Révolution', *R.H.D.E.S.* 1911, p. 153.

secondly, the agrarian distribution of the land between the cultivators, which rendered the fallow practically unavoidable. These
two facts, in turn, explain the nature of the cultivated plants and
also the peculiar structure of the rural society; finally they give
the key to the cattle situation and to the quality of agricultural
implements. The importance given in early agricultural treatises
such as Liger's *Nouvelle maison rustique* to the mechanical problems of ploughing, manuring, to the cultivation of corn and, to
a lesser extent, to the question of cattle helps us to understand
country life in old France. The question arises whether French
husbandry as it is expounded in such works was the same throughout the provinces. It was undoubtedly not entirely so. Indeed,
regional differences existed owing in some instances to a greater
development in the techniques, to a different partition of the soil,
or to geographical peculiarities; essentially in Flanders, in certain
districts of Lorraine and Alsace, in the vineyards of the South,
in the large mountainous areas of the Centre, the Alps or the
Pyrenees. But on the whole and for the greatest part of the
territory it is possible to speak of the 'ancienne agriculture', of
the 'ancienne manière' which was to be opposed by the 'nouveau
système'.[1]

The principal crop of this agriculture was corn.[2] There were
several reasons for its absolute predominance. Bread was the
principal food of the major part of the population, especially within
the agricultural world.[3] This fact explains the *a priori* importance
of all questions related to grain (export, exchanges between provinces, grain preservation, etc.). Furthermore, some kinds of grains
were very often used for animal as well as human consumption,

[1] Even in provinces (e.g. Normandy, Brittany) where enclosures already
existed and where therefore the agricultural landscape was different from that of
the open countries, the agricultural rhythm was no different.

[2] A. Young, *Travels*, vol. II, p. 616. 'Mémoire sur l'agriculture du Boulonnais
par M. de C.', *Bibliothèque Britannique* (Pictet), 1796, p. 150.

[3] A. Babeau, *La vie rurale dans l'ancienne France*, Paris, 1885, p. 88.

not only for pigs or poultry, but also for horses and oxen, which illustrates the acute shortage of feeding stuffs. The cropping system of a district always included a kind of corn which was sown in autumn. Often, too, varieties of spring corn, *les maïs*, were added together with beans and peas.[1] Besides corn, various industrial plants were also cultivated, such as madder, hemp, flax, etc. Vegetables, sweet corn, roots and artificial grasses like sainfoin or lucerne were cultivated round the houses, in gardens or in closes, but generally they played no part in the field system as a whole.[2]

Corn exhausts the soil more than any other crop, so that for an agriculture which still used archaic methods of cultivation, manures to fertilize the soil were absolutely indispensable. Indispensable therefore was the fallow which allowed the exhausted soil to rest. Thus, while part of the district was under cultivation, another part was lying more or less fallow, and the necessary food for the cattle was found in natural pastures, or on the verge of woods and forests and in common lands which were left uncultivated for that purpose.

If one bears in mind the well-known fact that the elongated shape of the holdings and their intricate texture of intermixed strips practically forbade any individualism in cultivation, and therefore compelled the rural inhabitants to a communal cultivation, one will be able to evoke the picture of an old French *terroir* with its clear and simple divisions. Here the village, its gardens and closes; there huge masses of green of different shades—winter corn, spring corn, the pastures and the fallow fields; yonder, the common lands and the woods; the only sign of a more individual cultivation being the *réserve* lands of the seigneur and the

---

[1] See Liger, *Oeconomie générale de la Campagne*, Paris, 1700, vol. I, pp. 338 ff.

[2] On local crops such as clover, rye as a green fodder, maize, etc. see G. Lefebvre, *Les paysans du Nord pendant la Révolution française*, Lille, 1924, p. 194; Duhamel, *Eléments d'Agriculture*, Paris, 1772; Valmont de Bomare: articles on *seigle, tréfle*, etc. in the *Dictionnaire d'Histoire Naturelle*.

park near the chateau.[1] So the rural countryside appears in the eighteenth-century *atlas-terriers* or in the illustrations of many agricultural books of the time which generally faithfully picture the rural scenery.[2]

This absolute necessity of a common cultivation was rendered even more unavoidable by the practice of *vaine pâture* and *parcours*, that is to say the common right of pasture over the stubble within the limits of one district or on the territory of a neighbouring one.[3] After the harvest all fields were open to the animals, cattle and sheep (pigs, goats and geese were not included in the right), which had hitherto been reared on hay or on what they could find in the common or on the fallow. But these supplies of feeding stuffs were not enough and the *vaine pâture* was a necessary addition to a too often meagre diet. As the *vaine pâture* lasted from August till February, no winter crop was therefore possible and all members of the community had to comply with the common agricultural rhythm.

Before showing the weak points of the system that was to be attacked by the agronomes, a word must however be said of its qualities.[4]

This agrarian organization and this cycle of cultivation maintained a strong community spirit among the rural society. Common works, common duties, common advantages too, attenuated to

[1] The seigneur's lands after the sixteenth- and seventeenth-century *réunions* were larger and better shaped than those of the *censiers*. See M. Bloch, *Les Caractères originaux...*, pp. 209 ff. There was no question yet of *grande culture* or *petite culture*. Indeed, large farms were not rare, especially on church lands, but as a rule the seigneur himself never exploited directly more than *quatre charrues*. The rest had to be let to farmers. In Liger, *La nouvelle maison rustique*, 1721, vol. II, p. 782.

[2] *Ibid.* vol. I, several engravings: *Labourage*, p. 519, *Moisson*, p. 547, *Récolte des foins*, p. 625. A suggestive description of this fact is given in G. Lizerand, *op. cit.* p. 108.

[3] On this question see M. Bloch, *Les caractères originaux...*, *passim.* Also Liger, Article: Pâtures, *La nouvelle maison rustique*, vol. I, pp. 778 ff.

[4] Lord Ernle (Prothero), *The pleasant land of France*, London, 1908, presents a remarkable understanding of the French agrarian structure.

34

a certain extent the differences between the rich and the poor in the same parish. Indeed there hardly could be any comparison between the wealthy *laboureur*, who possessed many fields and pastures and a large number of animals, and the wretched *manouvrier*, who possessed hardly anything. But the important fact is that the latter could survive. He often possessed in full propriety a cottage and a little garden, one or two cows which he kept on the common or added to the communal herd for the *vaine pâture*. And he felt himself a full member of the community. The moral advantages of a system which demanded solidarity in many domains have often been well pointed out.[1] The seigneurs themselves, or the *laboureurs*, or even the small peasants benefited in many cases from it. The former had important privileges as to the number of heads of cattle they could entertain with the common rights (*troupeau à part*) or else as regards grazing facilities (*herbes mortes*) in such regions as Lorraine, Franche Comté or Béarn. The latter also drew real benefits from the existence of these rights especially when they specialized in stock building or needed important quantities of manure for their lands. The fallow itself, later regarded as an *opprobre* to agriculture, could be rehabilitated and the old authors naively praised it, which provides an amusing contrast with the declamatory abuses of the later period. 'Un bon oeconome doit laisser reposer tous les ans le quart ou le tiers de ses terres alternativement; on épargne par la 1°) le tiers du bétail necessaire tant pour les labours que pour les charrois et autres choses 2°) les gages et la nourriture de plusieurs domestiques 3°) le tiers de la semence 4°) les fruits en sont meilleurs . . . . 7°) les travaux du labourage se font avec plus d'aisance et plus d'ordre.'[2]

Finally there was undoubtedly a real dignity in this old rural world which made the strong and powerful basis of the society

---

[1] Lizerand, *op. cit.* Compare his arguments with Ernle's: 'The open field system was proverbially the source of quarrels', *op. cit.* p. 156.

[2] Abbé de Petity, Article: Jachères, *Encyclopédie élémentaire*, Paris, 1730.

of the *ancien régime*. The *noble et antique manière* never lost its partisans.

Nevertheless, this system, the entire complexity of which has been studied by Marc Bloch, was opposed to an evolution towards agricultural improvements. It was essentially static, and perpetuated, together with its advantages, a great number of defects.

This common system of cropping, for instance, forced all members of the rural community to follow, not the most advanced and the most elaborate practice, but the slowest and the most backward. The general standard of cultivation, so as to be accepted by all had to be on the lowest plane. Therefore, an improved standard was almost unattainable. Though such crops as sainfoin and turnips were not entirely unknown, the *vaine pâture* system opposed itself to their extension on a large scale. Better methods of cultivation were forbidden because of the fallows with their heritage of weeds and thistles.[1] The yield remained very low, owing to conservative methods in tilling, selecting the seeds and sowing. Improvements in implements were almost impossible, so powerful was the weight of custom in this respect.

With the exception of a few large farming estates and of some provinces, cattle were primarily considered as manure producers, and therefore rather carelessly looked after. Selection and improvement of breeds, though sometimes advocated, was not followed. And the practice of herding together cattle and sheep on the common pastures led to every sort of disorder: the manure of the vagrant animals was practically lost for the land, diseases of all kinds were frequent and contamination easy and frequent and severe cattle plagues resulted. Finally, cultivated crops, and the borders of forests and woods naturally suffered through the unsupervised cattle's migrations.

[1] Irrigated meadows were also too frequently in a poor state. See 'Mémoire de Carlier', *Journal de Physique*, 1784; Pigeonneau et de Foville, *Administration de l'Agriculture au contrôle général des Finances*, Paris, 1882 (Année 1786); De Pradt, *op. cit.* vol. I, p. 142.

The practical facts besides were more or less codified in the theoretical works of the period. They all seemed to be the result of so many centuries of experience that farmers did not dare change anything to it. Liger, writing on ploughs, adds, 'Il est des charrues comme des terres; c'est à dire qu'où l'ancien usage veut qu'elles soient faites d'une telle manière, c'est abus d'y vouloir changer quelque chose'.[1] In the same fashion broadcast sowing, manure recipes, observance of astral influences, 'multiplication' of wheat thanks to marvellous 'elixirs', penning sheep in close and un-healthy *bergeries* was reverently maintained as the inheritance of a respectable tradition. This was the routine spirit against which later agriculturists were to exert their sarcasms.

Nous ne pouvons pas nous dissimuler que si notre système agricole est en général si peu convenable et mesquin, et si la France ne produit peut-être pas la moitié de ce qu'elle devrait produire, le génie national a apporté jusqu'à présent un obstacle invincible aux réformes de ce genre qu'on peut désirer. . . . En fait de culture surtout, nous sommes tentés de révoquer en doute ce qu'on nous apprend de nouveau et, le plus souvent, les cultivateurs ne savent témoigner leur reconnaissance à ceux qui cherchent à les instruire, que par des sarcasmes, des doutes et de l'incrédulité.[2]

Some good hints however can be found in those early agricul-tural treaties, but in a very humble way so to speak. Liger, for instance, warmly advertises the virtues of sainfoin and lucerne,[3] he shows a real concern for cattle and sheep, he alludes to the method of 'dérober la jachère', e.g. cultivation of the fallow with such crops as maize. Above all, the old treatises show a remark-able understanding of all questions concerning forests, trees, gardens and kitchen-gardens. But on the major issues the door

---

[1] Liger, *Oeconomie générale de la Campagne*, vol. I, p. 331.

[2] Article: Clôture, *Encyclopédie méthodique*. See also Fabroni, *Réflexions sur l'état actuel de l'agriculture*, Paris, 1780, in the Preface of which he contemptuously attacks 'routine'.

[3] Liger, *op. cit.* vols. I and II, *passim*.

was open to criticism and, after 1750, comparisons with the agriculture of England were to bring about demands for improvements.

The old rural world, however, was to resist firmly, facing the attacks of the new theories even under the pressure of economic factors of a general order. Indeed, the slightest change in this imperfect but strongly built system would have finally ended in a general upheaval, if the interests of all members of the rural society had been shaken and their relationships altered. In spite of the wretched condition of many peasants, there was a powerful factor of conservatism, fighting, for example, for the existence of a real peasant property or at least for the survival of the small *métayer* or *fermier* or even *manouvrier*.[1] In the rural society before 1750 the seigneurs were often not too exacting. Most of the *petite noblesse* still lived on its estates in a spirit of severe thrift.[2] In this case, too, conservatism often meant preservation of their privileges, of certain feudal dues exactly adapted to the old system of cultivation. Before 1750 there was in fact no urge to undertake great transformations in order to meet the economic changes and resist the economic pressure.[3] Until 1733 the only changes in the policy of the seigneurs were petty improvements in the levy of feudal taxes or increases in the rents after the renewal of farming leases. No general theory of how the agricultural structure of the country ought to be changed had yet been formulated. So that empiricism in economic matters within the old frame still prevailed. After 1733 the situation began to change. Sagnac has well pointed out what was the result of the economic evolution on French society: increase in the price of corn, increase in the rent of the land, increase in population, increase in the wealth of certain classes.

[1] See suggestive reflections by J. L and B. Hammond, *op. cit.* p. 105.

[2] Ch. de Ribbe, *Les Familles et la Société en France avant la Révolution*, Paris, 1874, p. 144. Pierre de Vaissière, *Gentilshommes campagnards de l'ancienne France*, Paris, 1903.

[3] Sagnac, *La société française sous l'Ancien Régime*, Paris, 1947, vol. II, which makes the greatest use of the conclusions of Labrousse.

After 1733 the *bourgeoisie* and the *noblesse de robe* began to invest money in the land, thus introducing into the countryside a new commercial spirit of profit-making, and often a taste and an interest in new methods.[1] The *laboureurs*, enriched by the high price of agricultural products, became more ambitious and more eager to acquire a profitable autonomy without suffering interference from smaller or poorer peasants. As for the landed nobility, provided that their old privileges be maintained, they would not oppose an attempt towards improvements and transformations which meant an increase of wealth. The provincial *moyenne* and *petite noblesse* began to supervise more closely the exploitation of their estates.[2] The new agricultural system advocated by the agronomes appeared to a great many as a new system of money-making.

But if there is a close connection between the economic pressure of the time and a necessary evolution of a static system of husbandry, it is nevertheless true to say that the connection is none the less closer between this attempt to change agrarian conditions in France and the publication of Duhamel's book.[3] Therefore, although this question of economic, agricultural and social transformations is a whole, the elements of which are intimately connected, it is nevertheless worth attempting to trace the evolution of the ideas expressed by Duhamel and to study their constitution into a whole agricultural system. In this respect it can be said that the *culture anglaise* as an organized theory was something absolutely new in France and this gives full value to the date 1750.[4]

---

[1] Sagnac, *op. cit.* vol. II, pp. 165-7.   [2] *Ibid.* vol. II, pp. 162-3.

[3] A discussion on the relative importance of social factors and technical discoveries will be found, in the case of England, in Hammond, *op. cit.* p. 36. The case is identical in France.

[4] It should be pointed out that all the demonstrations of the physiocratic school have a solid practical foundation which is generally found in the results of the *agriculture nouvelle*, as it was expounded by the agronomes.

CHAPTER IV

TULL IN FRANCE

JETHRO TULL'S theory, from the day it was formulated, has been the subject of controversies, extending to the present day. It is not intended to reproduce here the theory as it appears in the *Horse-hoeing Husbandry*. This has been done repeatedly[1] and has led both to high praise and vigorous criticism. Nor do we intend to make another critical study of the theory itself. There are grounds, as it seems, for Lord Ernle's views as well as for Mr Marshall's. But if Tull's work itself is not to be analysed, being absolutely outside our scope, its importance, in the light of the agricultural situation in mid-eighteenth-century France, cannot be too strongly emphasized.

Certain points are apparently agreed to be Tull's chief legacies to the new agricultural conceptions as a whole: clean farming, economy in seeding, drilling, the worthlessness of manures, and finally, the maxim that the more the irons are among the roots, the better for the crop.[2] While Lord Ernle praises him as the first to practise what is nowadays successfully applied at Rothamsted —i.e. a highly scientifically developed agriculture—Mr Marshall, besides criticizing his physiological theories, points out three major 'heresies': his opinions on the use of manures, his theory of perpetual cultivation of corn, and his advocation of a drastic

---

[1] See Lord Ernle, *English farming past and present*, London, 1936, p. 169; L. W. Moffit, *England on the Eve of the Industrial Revolution*, London, 1925; T. H. Marshall, 'J. Tull and the "New Husbandry" of the eighteenth century', *Economic History Review*, 1929, vol. II, pp. 41, 60; M. Bloch, 'La vie rurale, jadis et naguère', *A.H.E.S.* 1930, vol. I, p. 102.

[2] Lord Ernle, *op. cit.* p. 172.

reduction of seed employed per acre.[1] All this may be excellent for the purposes of English agriculture. That Tull's main ideas might have been either discovered or suspected before him, that his drill had been long known both in Spain and in some parts of England, is perfectly true. But in France, it was the first time that a scientifically constructed and complete agricultural theory had appeared. Hitherto, the names of Grew, Evelyn, Woodward, Bradley, etc., were almost completely unknown to the French public. So were their theories (which seem to have roused less interest than those of Tull, even in England itself). Other reformers may have been suspicious of Tull's 'sovereign treatment for all cases' in England. These other reformers were at the time unknown in France. Tull had the luck of being imported as the sole dispenser of an agricultural panacea at the time when theories coming from England were highly esteemed. It was merely his good fortune that, of all others, he it is who must be considered as the starting-point in France of new conceptions, as well as of the English influences, in agriculture.

The radical tone of his work, besides, must have helped him a great deal. Thus, he lets in daylight on the old empiric husbandry as it appears, more or less codified, in the treatises at the beginning of the eighteenth century. The scientific outlook of his work, together with its coherence, undoubtedly impressed his French readers, just as the new economic systems had done. No wonder, then, that its translation was entrusted to men of great repute in the French scientific world.

The vogue of his book for some time in England[2] among great landlords and members of the nobility also explains why it was introduced in France by people like the Maréchal de Noailles and the Duc d'Orléans.

Even if we agree with Mr Marshall that it did not start a kind

[1] Marshall, *op. cit.* p. 48.
[2] In Lord Ernle, *op. cit.* pp. 172 and 173, and Marshall, *op. cit.* p. 50.

of agricultural revolution in England, we cannot say the same for France. The judgment of a contemporary writer on the importance of his book on the destinies of French agriculture, is nearer the truth.

Ce n'est pas néanmoins que la mémoire de M. Tull ne doive être chère aux cultivateurs. Personne n'a plus contribué que lui aux progrès de l'Agriculture. C'est de lui que les fermiers ont appris à connaître les grands avantages qui résultent de la fréquence des labours qui divisent les molécules de la terre . . . qui en un mot fertilisent le sol et en plusieurs circonstances suppléent aux engrais. C'est lui qui leur a enseigné à tenir leurs terres nettes et exemptes de mauvaises herbes, à les préparer avantageusement à la production des grains en y semant d'abord de grosses raves et à bien tirer tout le parti possible de ces grosses raves en leur apprenant à les éclaircir par un labour à la houe. Ce fut M. Tull qui le premier introduisit l'usage de semer en plein champ, et par rangées, les fèves, les pois, les vesces, le sainfoin, la luzerne, et c'est aussi de lui que nous avons appris à ne plus prodiguer nos grains dans les semailles et à les répandres dans une juste proportion. C'est donc à ses lumières qu'est due la réforme qui s'est faite en Agriculture.[1]

Thus are the soundest principles of Tull's husbandry given their full value.

There is no doubt that had Duhamel only translated the *Horse-hoeing Husbandry* under the title of *L'agriculture par la Houe à Cheval* for instance, the book, interesting as it might have been, would have had the same fate as other English translations, and remained in the history of French agriculture only as of equal importance with Bradley's *Le Calendrier des Jardiniers* or Francis Home's *Les Principes de l'Agriculture*. The significance of Duhamel's book lies not only in the fact that he has conveyed Tull's main ideas, but in the fact also that he has assimilated them. The projected work, which was to be a mere translation, becomes

[1] *Voyage agronomique, précédé du parfait Fermier, ouvrage traduit de l'Anglais par M. de Fréville*, Paris, 1774, pp. xxxii-xxxiii.

in his hands an original work: Duhamel du Monceau's *Traité de la Culture des Terres, suivant les principes de M. Tull, Anglais.*

At the time Duhamel undertook the work, he had behind him a long career as agronome and specialist in things English.[1] He was already known as a specialist on wood-planting. After a journey in 1729 to England, where he had been experimenting with timber, he studied the action of rain on the quick growth of vegetables. In 1730 he produced a memoir on grafting,[2] in 1744 a book on layers and slips,[2] works of the same kind as Hill's *Timber-Tree Improver*,[3] or Stephen Hales's *Vegetable Statics*.[3] From Hales he was to borrow the main idea of the ventilation in his granary.[4] His *Traité de la Conservation des Grains* is almost contemporary with his adaptation of Tull. It may be said, then, that his treatise is not the first manifestation of English influence in French agricultural problems, but is an attempt at a reasoned analysis of certain agricultural tendencies from over the Channel, and an effort to adapt them to certain French methods. This view seems justified by the reactions to his work, anyway; for certain people accused him of wishing to submit French technique to a valueless foreign influence.

What is, then, the content and the meaning of Duhamel's treatise?

[1] There is no biography worthy of this eminent member of the French Academy of Sciences. The main information may be found in his 'Eloge' and that of his nephew Fougeroux de Bondaroy by Condorcet, and later on, some scattered details in the various agricultural periodicals of the beginning of the nineteenth century, when his memory was still venerated (*Mémoires d'Agriculture de la Société d'Agriculture du Département de la Seine*, and those of the *Société nationale d'Agriculture*). See also a rudimentary monograph by Drohojowska: *Les grands agriculteurs modernes*, Tours, 1885.

[2] See list of Duhamel's memoirs in Rozier's *Nouvelle table des articles contenus dans les volumes de l'Académie royale des Sciences de Paris, 1666-1770*, Paris, 1775.

[3] *La physique des arbres. . . .* Paris, 1758. This explains why this agronomer should have as a main title that of *Inspecteur de la Marine Royale* which had been conferred on him after his sojourn in England, when Buffon had been appointed, instead of him, as Director of the King's Garden. See *Annales du Muséum*, vol. IV, 1805, p. 19, n. 1.   [4] See below, ch. x, p. 159.

Tull's original work was in itself fairly voluminous inasmuch as it concerned soil cultivation only and was not a general treatise on farming. Its commentator enlarged it. The whole of the English agriculturist's doctrine is contained in the French work; in addition, all experiments, even the smaller ones, are quoted, discussed and elaborated. Duhamel adds his own views, and states his personal experience. He analyses and corrects. It is more than a 'good extract',[1] as the first volume consists of more than 400 pages, quarto. But, as this contemporary review of the book says: 'Mr Duhamel put into it some order . . . added some experiments in order to confirm the author's views.'[2] Doubtless, the intricate treatise of Tull required somewhat of a French dressing. After this first volume, Duhamel published five others, laden with experiments he and his correspondents had made, which conclusively showed how important the question was felt to be in the agronomic world. The content of the English book thus became a huge undertaking in six volumes, which excited the admiration of Tull's compatriots themselves.[3]

Volume I, which is the doctrinal and main part of the work, is divided into twenty-five chapters. The first five deal, as in Tull, with vegetable physiology, the function of leaves, roots, etc. They bear on the nourishment of plants, and Tull's experiments are reproduced. Duhamel, however, does not accept all the views of the English writer. After discussing the food of plants, he concludes,

Concluons donc qu'il est possible de se procurer tous les ans une bonne récolte de froment dans une même terre. Il ne faut pour cela que

[1] *Nouvelliste Oeconomique*, 1755, vol. IV, p. 4. Review of the *Traité*, in the Abbé de Petity's *Encyclopédie*, vol. I, p. 604, 'L'ouvrage tel qu'il parait aujourd'hui dans langue doit moins passer pour une traduction que pour un original.'

[2] 'Mr Du Hamel who dressed our great English husbandman, the celebrated Mr Jethro Tull, in French cloaths. . . .' *A Discourse on the Cultivation*, London, 1762, p. 4.

[3] See Preface to the English translation of *Practical Treatise of Husbandry, by the Celebrated M. Duhamel du Monceau*, London, 1759, p. 7.

multiplier les labours; que diviser suffisamment les molécules de terre; que mettre les Plantes en état d'aller ramasser dans la terre la nourriture qui leur est nécessaire; qu'empêcher les mauvaises herbes de la dérober à celles qu'on cultive et enfin de ne pas mettre dans un champ plus de plantes que la quantité qui y peuvent subsister. On satisfera à toutes ces conditions en adoptant la nouvelle façon de cultiver les terres.

The second part of the book deals with the treatment to be given to the earth in order to make it more productive (chapters VI-X). Duhamel has here a definite purpose in mind. He is far less interested in propounding a new theory than in uprooting certain fallacies from traditional agriculture. As a result, a lucid criticism of contemporary habits follows, in which he points out how badly cereals are cultivated. About those which he would like to see hoed during their germination he writes: 'La conduite de nos laboureurs est donc aussi peu raisonnée que si l'on donnait beaucoup de nourriture à un enfant et qu'on lui retranchât les aliments à mesure qu'il deviendrait plus grand'.[1]

Like Tull, he is a great supporter not only of preparatory ploughing, but also of preservation ploughing, during the growth of plants. But what proves (against his opponents) that he is not a blind follower of the English theory, is that he praises manures, whereas Tull condemns them. He agrees, indeed, that tillage is easier and more important in countries where fertilizers are scarce. He agrees that manure alters 'toujours un peu la qualité de la production'[2] but denies that it is harmful. Invoking 'l'expérience de tous les temps et de tous les lieux', he ended by entirely accepting the use of fertilizers. These are not only dungs. Duhamel, who in his *Eléments d'Agriculture* developed this question, recognizes some virtue in burning the grass of a field, provided that the peasants would not confine themselves to this rather primitive practice. He thus advises *essartage* and *écobuage*,[3] 'comme je l'ai vu pratiquer en Bretagne et dans d'autres Provinces du Royaume.

[1] *Traité*, ch. IX, p. 176.　　[2] *Ibid.* Préface, p. xxvi.　　[3] Paring.

M. Tull désapprouve cet usage; il est néanmoins d'expérience que par cette pratique, on communique aux terres une fertilité qui dure plusieurs années.'[1]

It is therefore certain that Duhamel's position is less uncompromising, less systematic, more moderate than Tull's, and that he attempts to retain any of the French methods which he thinks worth preserving. A good example of this moderation is the criticism he makes of Tull's method of cultivation in ridges (*billons*), and he gives instances from the Loire valley where the practice is used and does not give the results claimed by the English author.[2]

Tull's four-coultered plough also does not much please him. Indeed, it may be useful in breaking up and tilling strong lands; 'mais notre Auteur', says he, 'ne bannit point la charrue ordinaire; il en approuve même l'usage autant que je le puis croire, pour les labours d'Été.'[3] But in the end, he prefers the traditional ploughs, 'qui n'exigent pas autant de chevaux, la dépense revient à peu près au même et la terre est brisée et émiettée au lieu d'être renversée par grands gâteaux'.[4] On the other hand, he strongly advises the horse-hoe which he calls 'charrure légère de M. Tull', which he thinks very handy and useful for ploughing in between the rows. The roller, as improved by the English author, he keeps in his new method, although, anxious that the earth might be too compressed, he advises that it should not be used after rain.

---

[1] *Traité*, vol. I, Préface, p. xxviii.

[2] In particular cases, however, he approves of this method and gives an interesting instance of the marshy lands of Gâtinais: 'Les terres du Haut Gâtinais retiennent l'eau, ce qui oblige les Fermiers de labourer par planches et même de n'ensemencer que la partie la plus élevée des planches, de sorte que plus de la moitié de la terre reste vide de grains presque vers le milieu du Printemps; alors les eaux étant retirées, et la terre desséchée, ils labourent à bras cette terre qui n'était point ensemencée et ils sèment des navets, des fèves, des pois etc. . . . Leur intention se borne à la vérité à tirer un profit de toute leur terre; mais il est certain que le grain qui avait été semé sur le milieu des planches tire un avantage sensible de ces labours.' *Traité*, Préface.

[3] *Ibid.* ch. VIII, p. 111.　　　　　　　　　　　　　[4] *Ibid.* ch. XXII, p. 329.

The question of the Drill, one of the main points of the English technique, called for a lengthy explanation. In the edition of his work, Duhamel's description of Tull's machine appeared to his contemporary readers too complex and intricate to understand, and he abandoned it in the following editions, confining himself to a minute explanation only of the copper-plates which made up his own drill.

It was less the idea of a minute and improved cultivation, than that of a mechanized one, which was to condemn Duhamel's husbandry to exist in the eighteenth century only as an interesting forecast of what agriculture was to become a century after him. It is, in this respect, to modern agriculture, what Cugnot's steam-machine is to our modern railway-engines. Its documentary value at any rate, is very great.

Finally, Duhamel directly challenges, as his model did, the principles of traditional husbandry. The method of sowing in rows with beds and ridges in between, and with frequent hoeing, is elaborately described. He compares in great detail the 'méthode ordinaire' and the 'nouvelle méthode' and shows all the advantages of the latter.[1] He therefore emphasizes the necessity of giving up the fallow system. There is little doubt that in 1750 he thinks a perpetual cultivation of the same crop possible, in the case of wheat above all.

Tull's work also afforded Duhamel an opportunity to devote important chapters to new crops, the cultivation of which was still scarcely known, namely, turnips, sainfoin and lucerne.[2] These were not absolutely unknown in France. They are quoted with much praise in the old treatises of Liébaut and Serres. Liger held sainfoin in high esteem. But Duhamel's insistence on them is significant. He is no doubt the first man in France to have systematically shown the advantages of suppressing the fallow system, thanks to a better understanding, and extension of artificial

[1] *Ibid.* vol. I, ch. XXI, pp. 297-301.　　　[2] *Ibid.* chs. XIII, XIV, XIX, XX.

pastures. These fodders, he says, allow a more highly developed breed of cattle, whose manure is used for improving the lands.

Turnips, it seems, were at the time commonly known only in Central France, Auvergne and Limousin. Hitherto they were often called by their provincial name of 'rabioule'[1] and it is only around 1750 that they acquire their significant name of 'turnep'.[2] By the new method of cultivation they were to be grown much bigger, and so the profit was to be greater. They were particularly useful for feeding cattle. He then discusses the way they should be used. He opens also the question of keeping sheep in a field of turnips, a question to which the greatest importance seems to have been attached by the agronomes, for it is much spoken of in many of the agricultural books of the time, and was, as we shall see, debated all through the century. Furthermore, he praises sainfoin and lucerne and describes the qualities of artificial meadows. According to him, we must impute to backward routine that fact that 'chaque Province qui est en possession de cultiver certaines plantes, ne songe nullement à tenter la culture de celles qu'on élève avec avantage dans d'autres Provinces. J'ai vu le sainfoin absolument inconnu dans les provinces où la terre me paraissait très propre à en produire. On ne sème en France le fromental,[3] dont les Anglais font un cas singulier, que dans quelques cantons aux environs de Lyon.'[4] The peasants' ignorance about artificial meadows draws from him a sigh similar to Arthur Young's. 'Je vois toujours avec regret des friches immenses uniquement réservées pour nourrir une petite quantité de bétail, pendant que la dixième partie de ce terrain, bien entretenu en prés artificiels pourrait produire une bien plus grande quantité de fourrage.'[5] Finally, the subject of the cultivation of wheat leads him to give a more precise

[1] Pigeonneau et de Foville, *op. cit.* p. 384; *Bibliothèque physico-économique,* 1786, vol. II, pp. 109 ff.

[2] Often called *navet anglais.*

[3] On the confusion between 'fromental' and 'ray grass' see below, ch. VII.

[4] *Traité,* vol. v, Préface, p. xviii.   [5] *Traité,* vol. v, Préface, p. xxvi.

description than Tull's of its diseases. Although substantially developed, this part of agriculture was to become the field of his colleague Tillet, whose active researches on this question are well worth mentioning.[1] Duhamel, however, had shown the way in this also.

The significance of this first volume of Duhamel's *Traité* is very great. Before 1750 agricultural literature was far less important in France than in England.[2] Afterwards, the number of volumes on agricultural questions was to increase enormously until it constituted a collection which has hitherto never been systematically studied. This important number of technical writings is to be attributed to the fact that with Duhamel, agriculture became a science. Prior to the Revolution, the major part of these writings was in fact devoid of any elaborate political conceptions. H. Sée justly points out: 'Do not let us forget the surge of an agronomic literature even before the beginning of the physiocratic movement.'[3] The goal of agriculture must of course be France's wealth, but its immediate interest was scientific, just as was that of physics or natural history. It is Duhamel's glory that he led the way. The criticism and praise of his work shows that he had captured the public. There is scarcely any work of value belonging to this time which does not to some extent show its writer to have been well aware of the new theory. Duhamel has a right, then, to be considered a pioneer of French agriculture and among the pioneers of modern agriculture generally. This title he owes to the English model, which he was the first to introduce to

[1] Especially those undertaken at Trianon (*Précis des expériences faites à Trianon, sur la cause qui corrompt les blés*, 1756) and in Angoumois where he studied a corn pest on a large scale (*Histoire d'un insecte qui dévore les grains dans l'Angoumois*, 1763).

[2] Activity in this particular field in England is well shown in G. E. Fussell's book, *The Old English Farming Books from Fitzherbert to Tull—1523 to 1730*, London, 1947.

[3] *Histoire économique de la France*, with notes by R. Schnerb, Paris, 1939, p. 202, n. 3.

France. This fact has been noticed, though timidly stated, by historians of agriculture. Morogues wondered in the eighteenth century why Loudon had not developed in his book[1] 'the history of the influence of English agriculture on that of France, an influence so strongly marked in Duhamel du Monceau's writings in the middle of the last century'.[2] A more recent historian also notes: 'it is no accident that Anglomania and agromania occurred at the same time'.[3]

What incentive lay behind Duhamel's aim and what was the result of his work?

He was passionately interested in Tull's theory. For years he experimented on it on his own estates and extensively published his results and those of his followers. He examined it minutely, chiefly because he had long shared in its field of experiments, and also because it enabled him to verify his own discoveries. This explains the particular care with which he discussed Tull's experiments in vegetable physiology. Very often he does not give his own views. He gives a full account only of the known discoveries, of Tull's ideas, of his own experiences; he refuses sometimes to come to a conclusion. Still, his work is like a milestone marking the progress botany had made and the road that still remained.

His main concern is the testing of a new agricultural technique. The chapters he devotes to the description of traditional cultivation are certainly among the most precise pictures we possess of the technique of soil cultivation in France under the *ancien régime*. He had travelled and had actually seen the state of agriculture in the provinces, giving, for instance, precise descriptions of the methods used in Brittany, Gatinais, Lyonnais and in his own estate near Pithiviers. He praises here, criticizes there. On the whole he considers the technique backward and careless. He often

[1] Loudon, *An Encyclopaedia of Agriculture*, London, 1825.

[2] *Cours d'Agriculture*, Paris, 1840, p. 88, n. 4.

[3] N. S. B. Gras, *A History of Agriculture in Europe and America*, New York, 1925.

speaks of the routine and lamentable tradition of the French peasants. That is why his work essentially tends to the description of a 'new method', born in a country whose agriculture is becoming famous; a minute and complicated agriculture, but with assured advantages.

We saw earlier, in attempting to analyse his book, what this system of agriculture implied:

(1) Careful tillage of the soil and careful cultivation;

(2) Saving of seed and maximum produce;

(3) Modification of the triennial system with fallow into scientific rotation of crop.

(4) Importance of artificial fodder and of bringing little-used pastures into cultivation.

(5) Improvement and modernization of agricultural implements, of harvesting processes, and of the storage of the crops.

In short, the requirements of any agriculture at a superior stage.

All these things, which were beginning to be known in England when he translated Tull, and which were to develop as the century went on, he presents to his fellow countrymen. This, as it seems, is his most immediate purpose. His teaching was elaborated over a period of ten years. Naturally, in the *Traité*, the new theory, because it is new, is more drastic than in the *Eléments*. That is why it would be a mistake to judge Duhamel from one only of these works. For between writing the two he had readjusted his point of view towards a happier combination of the French and English systems. The mistake however, is often, almost always, made.[1] That is also why he is less of a translator than an adapter. He is well aware of the reactions which must follow his undertaking. Prejudice will be difficult to extirpate. Thus he desires some moderation in the application of his method. The peasant must

---

[1] When M. Bloch dates the introduction into France of English methods at 1760, year of the publication of the *Eléments d'Agriculture* (*Les Caractères originaux*, p. 221), he seems to ignore the opening of the debate 10 years before. This was, however, pointed out in Wolters, *op. cit.* p. 159, n. 1.

be converted through example. He defines the practical way to follow and the duty incumbent upon the leading agricultural class thus:

Car, qu'on ne s'y trompe pas; les choses trop recherchées et compliquées ne conviennent point aux objets très étendus; et c'est dans ce cas que se trouve l'Agriculture. Une culture très savante qui exige des soins et des attentions particulières et scrupuleusement suivies pourra réussir dans un terrain d'une petite étendue et placé sous les yeux d'un propriétaire assidu et intelligent, mais entre les mains du commun des cultivateurs, des opérations mal combinées produisent des effets tout opposés aux vues qu'on s'était proposées. Je pense que bien des gens instruits doivent essayer de détruire les préjugés et les routines reconnues vicieuses; mais je leur conseille de ne les attaquer que peu à peu et de se garder de vouloir changer brusquement les usages établis depuis si longtemps dans leurs cantons.[1]

[1] *Eléments d'Agriculture,* Préface, vol. I, p. vi.

CHAPTER V

CONTROVERSY ON DUHAMEL'S
NOUVEAU SYSTÈME

THE new method of Tull and Duhamel had an extraordinary vogue in France. Until the end of the century in such *Summae* as treatises by Rozier[1] and Tessier,[2] it completely dominates the agricultural question. Wherever corn cultivation is under discussion, even though the writer may sometimes be of a different opinion, the names of Duhamel and Tull appear. The argument was at one time passionate. Books were printed attacking the new theory, others supporting it; on the whole, the 'new system' attracted the interest of the agricultural world.

The study of this controversy brings to light a great many shades of opinion. However, whether attacked or supported, the importance of the Tullian system does not lie in its contemporary success, either as pure theory or as a practicable measure. Tull's husbandry was not England's husbandry in the eighteenth century, and it was never that of France, apart from a restricted circle of experimentalists or *curieux d'agriculture* (agricultural connoisseurs).[3] Its intrinsic agricultural value might be approved or condemned. But even had it never been applied in practice, it would still have great historical significance—just as the theory of Phlogiston has historical value in the study of chemistry. It gave rise, in France, to a lasting discussion featuring prominently in the history of scientific ideas of the eighteenth century.

Another noteworthy thing about it is that it is responsible for the conception of later agricultural works, as much so—if indeed

---

[1] *Cours complet d'Agriculture . . . ou Dictionnaire universel d'Agriculture*, Paris, 1781. [2] *Encyclopédie méthodique*, Section: Agriculture, 1787-1821.
[3] See below, ch. XII.

not more—as the physiocratic movement.[1] It is also well worthy
of note that the considerable number of volumes, memoirs or
articles following Duhamel's treatise, seem to owe their existence
only to the interest it evoked. Reactions which followed it were
almost exclusively of a scientific nature.

It may seem, and this is an argument often met with, that after
all, Duhamel's method is not new; that the whole of his research
can be replaced within the general evolution of agricultural
technique all over the world. In the remote past, the Chinese had
tried to solve the problem of seeding with the help of a special
instrument, the Drill. The most complex machinery in modern
North America must be but the outcome of a technique striving
towards fulfilment throughout history, and in which the endeavours
of Tull and Duhamel are but an episode.[2] The charge that their
work was unoriginal was immediately brought against them,
inasmuch as Duhamel had scrupulously mentioned a similar invention which appeared in Spain in the eighteenth century.[3] True,
before Tull and Duhamel, the 'new method' had been tried
sporadically and locally.[4] But the fact is that Baddam's article in
the Philosophical Transactions of the Royal Society,[5] although it

[1] At a time when the physiocrats were forgotten or at least discredited, the
names of the pioneers of French agriculture were still remembered by their
successors. 'Si l'Angleterre peut se vanter d'avoir un Bakewell, un Coke, un Ellman,
un Sinclair, un Sommerville . . . n'avons nous pas aussi nos Parmentier, nos
Tessier, nos Despère et s'ils ont eu leur Miller, n'avons nous pas eu aussi notre
Duhamel?' Mémoires d'Agriculture (Report of Yvart on English Agriculture,
vol. x, p. 93).

[2] Article: Agriculture, Grande Encyclopédie, Paris, n.d. Chinese agricultural
technique was expounded in the reports of missionaries. These often emphasized
the kind of religious aspect of agriculture in China and its complexity. See
Histoire générale des voyages, Paris, 1761. Table des matières, vol. xvi, Article:
Agriculture. See also, Practical treatise of husbandry . . . by the Celebrated
M. Duhamel du Monceau, London, 1759, ch. iii, p. 306.

[3] Traité, vol. i, ch. xxv, p. 373. On this point, see a curious and minute mise
au point by Russel H. Anderson in 'Grain Drills through thirty-nine centuries',
Agricultural History, 1936, p. 166.

[4] Tull's enemies accused him of having used the invention of Plat, Platter or
Worlidge.   [5] No. 60, p. 1056.

pointed out this method, had found not the slightest response at the time.[1] On the contrary, it may be said that, when Duhamel's work appeared, the method was unknown to French agriculture and was criticized as an innovation.

As soon as it was published the *Traité de la Culture des Terres* found in France a very extensive public.[2] Undoubtedly the work attracted its French readers chiefly by the novelty of its form and content. It reached individuals, already interested in those questions, who moved in circles where English influence had, to some extent, penetrated. Immediately before the publication of the treatise there were already many who had thrown themselves into agricultural experiments. The publication of the treatise brought to the fore hitherto unknown names from all parts of the kingdom, from the most remote and backward provinces, beyond the border even. The book of the French agronomer came at the right moment.[3]

Duhamel himself keeps us informed of the fate of his book among agricultural amateurs. His recognized honesty and loyalty alone allow us to take for granted what he says about a spreading taste for agricultural experiments due to his book. 'We cannot conceal our satisfaction at seeing lovers of agriculture multiply. We know them to be in every province of the Kingdom.'[4] This fact is confirmed by all the accounts we find in the contemporary periodicals, and the strongest evidence is afforded by the five volumes published after the first doctrinal one, from 1751 to 1756, where we have accounts of results sent by Duhamel's correspondents every year, with the author's comments. In fact, every

[1] It seems, however, established that Tull knew it; on this point see *Dictionnaire d'Agriculture . . . de l'Institut*, vol. II, p. 457.

[2] 'La méthode de cultiver les terres d'après les principes de M. Tull, proposée par M. Duhamel du Monceau . . . acquiert, Monsieur, tous les jours de nouveaux partisans', *Année Littéraire*, 1755, vol. VIII, p. 262.

[3] Patullo wrote about the volumes of the *Traité*: 'on n'en connait en aucune langue d'aussi bien faits et qui aillent si parfaitement au but.'

[4] *Traité*, vol. IV, Préface, p. i.

province, every geographical region, seems to have possessed at this time a great landowner conscious of the reforms that must be brought into traditional farming and who, therefore, carried out experiments. Far-seeing and progressive landlords were not lacking; nor was help from village parsons and big farmers.[1] It is here that we may realize the overwhelming weight of economic and social conditions that lay on an agricultural technique which, in the light of the new experiments, would willingly be transformed.[2] The great difference between England and France is testified to by the fact that, whereas in the former country the work of a Townshend produced almost immediate results,[3] in the latter, although (and this is an important point) the Townshends were not lacking, their efforts were not to bear fruit until a century later, after the Revolution.

However, between 1751 and 1756, Duhamel received regular information about the success of his new system. He corresponded with agronomes whose names are forgotten now, MM. Harrouard of La Rochelle, Baron de Meslai, Van Dussel of Bayonne, Bonnet (F.R.S.), Navarre, Dean of the Bordeaux Cour des Aides, the Presidente d'Augeard at Torgan, Eyma and Boissière of Bergerac, de Vormesel, gentleman of Périgord, etc.[4] It is worth noting that most of these names are grouped in the South-western part of France, round Bordeaux, possibly because of the activity displayed by the Academy of Bordeaux. (Tillet's

---

[1] *Instructions de Morale, d'Agriculture et d'Économie . . . ouvrage destiné à servir pour enseigner à lire aux enfants de la Campagne*, Abbé Froger, 1759. *Boussole agronomique ou Guide des Laboureurs . . . traduit du Latin par quatre curés de Normandie*, Paris, 1762. On an *Ecole d'agriculture* organized by a *curé du Maconnais* in his vicarage, see *Journal de Physique*, Introduction, vol. I, 'Instructions d'Agriculture'. Also *Prix de l'Académie de Soissons*, 1778, etc. On the role of the *curés* see Babeau, *Le Village sous l'Ancien Régime*, Paris, 1878, p. 129. Also P. de Vaissière, *Les Curés de Campagne au XVIIIe siècle*, Paris, 1933.

[2] According to *Le bon fermier ou l'Ami des Laboureurs*, 1767, the peasant would be prepared to perform experiments, but, 'c'est le défaut d'aisance qui l'empêche de faire des progrès', *Année Littéraire*, 1787, vol. III, p. 305.

[3] Ernle, *op. cit.* p. 175.   [4] *Traité*, vol. II, *passim*.

Memoir on corn diseases,[1] for instance, was to be awarded a prize —a proof that particular interest was taken in problems about corn in that part of France.) We hear, besides, the name of M. de Gourgues, Président à Mortier at the Bordeaux Parliament, who was the first, with the Duc d'Orléans, to bring one of Tull's drills into France.[2] These facts show without possible doubt that between 1750 and 1760 an enlightened movement of agricultural experiments was taking place in Bordelais. Duhamel supporters are also to be found in the Comtat, in Normandy, in Lorraine, while in Switzerland, near the border, was Lullin de Châteauvieux whose links with France[3] and the publication of whose books in Lyon, allow us to consider him as a French agronome.

Results, according to these writers, seem to have been extremely encouraging, with the natural exception of years of particular drought. Some of them applied Duhamel's and Tull's principles literally, and deprived the crop of all manures,[4] submitting it instead to frequent and minute ploughing. But this extreme practice did not last long. Most generally, manures as well as ploughing were used. The principle of the new method was retained however; plants further apart from each other, cultivation in rows and ploughing in between. These experiments proved successful.

M. Eyma me marque qu'il a été très satisfait de sa récolte. M. de Brue a eu encore un plus grand sujet de satisfaction. Un gentilhomme du Poictou, très zélé pour la perfection de la culture des Terres m'écrit que la récolte du champ qu'il a cultivé suivant nos principes a été aux

---

[1] *Essai sur la cause qui corrompt et noircit les grains dans les épis*, Bordeaux, 1755, vol. II.

[2] *Traité*, vol. II, p. 57. Points out the activity in agricultural research round Bordeaux; *ibid.* p. 87.

[3] Syndic of the Republic of Geneva. His son was colonel of the Swiss regiment de Châteauvieux, the rebellion of which was one of the first episodes of the Revolution. Lullin was always referred to as a French agronome.

[4] Report of La Michodière, Intendant of Lyon, in *Mémoire sur la pratique du semoir*, Lyon, 1761, by Châteauvieux, which certifies the success of the undertaking.

meilleures récoltes faites dans ce même champ comme sept est à quatre. ... Dom Edouard Provenchère, Procureur de la Chartreuse du Liget près Loches m'informe qu'une épreuve qu'il a faite sur l'orge l'a beaucoup satisfait. M. de Châteauvieux continue d'être content de ses récoltes.[1]

It would be tiresome to give here all the details of these testimonies. Be it noted, however, that the pages dealing with these experiments are most interesting agricultural documents. The least details of soil cultivation are studied with the cost price of each, its adaptation to the nature of the ground, climate, etc. These accounts are most revealing as a whole. Duhamel knew that only repeated and successful experiments could convert people to scientific husbandry.[2] As the publication of his observations went on, the number of his correspondents increased. He remained the centre of an activity which henceforth was no longer his, as the *Année Littéraire* points out: 'Vous voyez, Monsieur, par les différentes personnes qui ont eu part à ce traité, que les recherches qui tendent à perfectionner l'Agriculture ne sont plus l'ouvrage seul de M. Duhamel; elles sont devenues un travail commun et il s'est formé une espèce d'Académie dont les sujets sont répandus dans tout le Royaume.'[3]

It is then certain that the joint conclusions of Tull and Duhamel aroused, after their publication in France, very wide interest; that experiments were made according to their principles and finally that they enjoyed great fame among certain members of the agronomic movement. The new method indeed was never popularized, but it was broadcast widely enough to allow the inference that it reached the majority of individuals capable of being interested in it.

---

[1] *Traité*, Préface, pp. vi, vii, vols. IV, V.

[2] 'The 4th and 5th volumes of M. Duhamel's work, which contain the greatest part of the available experiments ... deserve still higher commendation, and may yet more justly be proposed as models, not only for their accuracy and success.' Duhamel du Monceau, *Practical Treatise of Husbandry*, Préface, p. vi.

[3] *Année Littéraire*, 1755, vol. VIII, p. 269.

This interest, nevertheless, did not show itself only by praises. Various arguments were brought forward to contradict Duhamel's conclusions, which were sometimes violently attacked.

First of all, the new method, considered as a pure innovation, upset the peasant's routine and traditions. This idea was constantly referred to at the time. Châteauvieux speaks of the peasants' 'repugnance to accept new practices'. The system was called the dream of a man locked in his study, a dream 'that the peasant grasps with great difficulty. There ought to be something clearer, more precise, more fixed'.[1] Supporters of the treatise themselves objected to its unnecessary length; its adversaries took up this argument also. Desplaces says ironically: 'There are a great many excellent, though very well-known, details in the *Traité de la Culture des Terres* which has just followed up to its sixth volume; it is not likely to have ever been read by any farmer in the kingdom.'[2] Another says the same: 'Among the readers themselves, very few are cultivators and it is surely not the farmer of the Brie who will occupy his leisure in wading through M. Duhamel's volumes.'[3] Opponents of the new system pretended to think that the 'new husbandry' as conceived by its author, should be immediately intelligible to the French peasant world and rapidly broadcast. Duhamel has answered this objection himself.[4]

There were other reasons for opposing the new system. Apart from apathy, there may have been the discouragement of people who had tried and not succeeded. To them, novelty was synonymous with failure. This tendency is well expressed in these lines:

I am expecting many contradictions from two sorts of persons, absolutely opposed in their way of contemplating the problem. The first are really the 'people', who have given up the privilege of thinking

---

[1] Prix d'Agriculture, *Année Littéraire*, 1764, vol. I, p. 312.
[2] *Préservatif contre l'Agromanie*, Paris, 1762, p. 73.
[3] *Ecole d'Agriculture*, Pamphlet in 12 Estienne, Paris, 1759.
[4] In the preface of the *Eléments d'Agriculture*, p. vi.

and submit to tradition. With them, discussion is impossible. The others have for many reasons, left the traditional way. They have thought and experimented but either having failed, they mistrust what is new or their prejudice does not allow them to learn anything about an art in which they think they have made particular progress; they are none the less full of prepossessions against this system.[1]

This is a very judicious remark. It is astonishing, when reading the agricultural literature of the eighteenth century to notice so much fatuity and vanity in some authors, as though they were writing about one of the great philosophical or artistic controversies of the century. It is indubitable that Duhamel's work has been attacked on purely invidious and jealous grounds.

The most interesting criticisms, however, are those of impartial writers, who were really concerned with the technical aspects of the work. All the experiments in connection with the new cultivation were not equally successful. Here and there, we hear echoes of those country Assemblies where the vicar or some farmer explained the new system, and gave advice which sometimes ended in failure. We find that in a 'Lettre écrite par M. de . . . à l'éditeur de la *Nouvelle Culture des Terres* de Duhamel du Monceau':

Let me tell you about the progress your paper makes. Since I have been receiving it regularly, a kind of small Academy of Agriculture has established itself at my house; it means that my neighbours, keen on everything related to this Art, gather regularly in my house as soon as I receive my copy. We read it and examine its contents together. I bought two years ago the *Nouvelle Culture des Terres* by a celebrated Academician in whose favour I was at first prejudiced. The title of Academician helps, as you know, to give a favourable impression. Who can deny it? Besides his system derives from an Englishman . . . it is really enough to make a semi-scientist believe in it as an article of faith; thus it was only after many an argument and reasoning for and against with my rustic Academicians, that I decided to give up my

---

[1] 'Nouveau système d'Agriculture', *Nouvelliste Oeconomique*, 1754, vol. I, p. 87.

mistaken prejudice; they demonstrated with obvious reason that the English system was an attractive idle fancy, and nothing more.[1]

Criticisms afterwards become more precise. Tull's ploughs and drills were declared unpractical and of a very imperfect action. Also they criticized the loss of ground caused by cultivation in beds, a point on which however, Duhamel had strongly insisted, showing that the loss of arable land was largely compensated for by a greater abundance of the crop.[2]

The general tone of these criticisms must be taken into consideration. It shows firstly a sort of contempt for scientific agriculturists, or at least some distrust of the lights of the Academy when it is a question of practical matters.[3] Also the hint about an 'English method' indicates a reaction against the influence of England which began to be felt in France in a way that some thought excessive.[4]

All these objections were more impressively concentrated in a pamphlet, the provoking tone of which was pointed out in contemporary reviews.[5] This work, whose title *Préservatif contre*

---

[1] *Jour. Oecon.* 1753.

[2] *Traité*, vol. I, ch. XXI, p. 284; *Eléments*, vol. I, pp. 485-97.

[3] Marquis de Mirabeau wrote: 'l'intervention scientifique aux chose usuelles est souvent dangereuse en ce que le tic des savants est la découverte; qu'une prétendue découverte entraîne tout aussitôt l'anathème sur tout usage contraire ou qui ne dérive pas de ce nouveau principe; d'où résulte opposition contre les spéculateurs et les agents, et consequemment danger de l'autorité dans les mains des uns et des autres.' *L'Ami des Hommes*, La Haye, 1759, vol. IV, p. 382.

[4] Extremely noticeable in the second half of the eighteenth century, mainly because of political motives, this reaction was particularly strong during the Empire and almost systematic. Desplaces is obviously anglophobe: 'Il devient fort à la mode de prendre en tout les Anglais pour modèles. . . . Leurs écrits rustiques, traduits ou commentés en notre langue n'annoncent pas la supériorité qu'on veut leur accorder, etc.' *Histoire de l'Agriculture ancienne*, Paris, 1765, Préface, p. xvii. Same tendency in de Planazu, 'Je n'entrerai à cet égard dans aucun des parallèles qui souvent conduisent aux plus grossières erreurs; tel est le parallèle que l'on voulait que je fisse avec l'Angleterre.' Same in Carlier, 'Ces derniers traits et une partie de ceux qui précèdent, devraient bien nous mettre en garde contre les exagérations de la forfanterie Anglaise: elles font trop de dupes parmi nous.' *Observations historiques*, p. 279.

[5] *Jour. Oecon.* 1762; *Année Littéraire*, 1762.

l'Agromanie[1] can be compared with the no less suggestive *Préservatif contre l'Anglomanie* of Monbron, develops arguments raised both against the new method and against its English origin. The enthusiasm with which all the agronomic writers celebrated English techniques and the lessons taken from their improved methods, seemed to annoy this fanatically French agronomer who thought the peasant's customs good enough, provided they are moderately corrected.

The fundamental criticisms (as the book is written by a very able and sensible man)[2] are worth mentioning. He writes: 'Cette méthode . . . a eu moins de partisans en Angleterre où elle est née, qu'en France où elle n'est qu'adoptive. . . . Il m'a paru que cette culture avait un vice intérieur que rien ne pourrait jamais corriger.'[3] The most obvious drawback is the awkwardness of the machinery. He writes about the drill and after many a joke about the complexity of Duhamel's description in the first edition of the *Traité* he points out its defects. 'Il y a sans doute une fatalité singulière attachée à cette machine.'[4] Tull's ploughs also are wretchedly built. They have been, after all, obliged to give them up and finally go back to the traditional ones.

'On a tout prévu; il fallait une charrue particulière pour exécuter les labours qu'il faut donner aux plates-bandes; pour cet effet, M. Tull a imaginé une Houe à chevaux; c'est une manière de petite charrue sans roues; on l'a perfectionnée en y ajoutant d'abord une roue, ensuite deux, puis un soc; on en a aussi inventé une autre qui remue seulement la Terre en dessous sans déplacer sa superficie. . . .'[5] Besides this, the new agriculture had some grave

---

[1] Desplaces, *Préservatif contre l'Agromanie ou l'Agriculture réduite à ses vrais principes*, Paris, 1762.

[2] F. Brunot's criticism of the book (*Histoire de la langue française*, p. 197) seems unjustified. It contains on the contrary very moderate views about Agriculture, but firmly supports well-understood tradition.

[3] Desplaces, *Préservatif contre l'Agromanie*, p. 72.

[4] *Ibid.* p. 151.   [5] *Ibid.* p. 152.

defects: 'Les labours fréquents prescrits aux rangées de Bléd, leur nuisent peut être dans tous les temps, et certainement dans les années sèches.'[1] Being opposed to too frequent ploughing, he advocates the traditional use of manures and maliciously points out that Duhamel himself in the *Eléments d'Agriculture* dares not prohibit them. Rotation of crop he does not like either, any more than artificial meadows. Desplaces' position is curious because of its exaggeration. In 1762, the great use of artificial meadows had been amply demonstrated by many a serious study and experiment.[2] However, after a violent attack against this method of farming in general, he concludes: 'Les Prairies artificielles ne peuvent guère servir qu'à la nourriture des bêtes de tirage et à celle d'un petit nombre de vaches . . . en un mot, les Prairies artificielles ne sont point à comparer aux naturelles.'[3]

This is a particularly violent form of the reaction against innovation in traditional husbandry. His views are decisive: 'on ne saurait donc mettre la méthode de M. Tull au nombre des découvertes utiles.'[4]

Another element in the controversy, a rather secondary but serious one, is the tone of approbation of the Royal Censor.[5]

La Salle de l'Etang holds the same opinion as Desplaces. But

---

[1] *Ibid.* p. 159. The argument will be taken up again by Rozier, Article: Amendements, *Dictionnaire Universel*.

[2] Especially in La Salle de l'Etang's *Prairies artificielles ou Lettres à M. de...*, Paris, 1756. Other books mentioned in ch. VII.

[3] *Préservatif contre l'Agromanie*, p. 117.

[4] *Ibid.* p. 162.

[5] The complete text of the approbation appears as follows: 'J'ai lu par ordre de Mgr. le Chancelier un manuscrit qui a pour titre: l'Agriculture réduite à ses vrais principes. Les vérités simples qui y sont exposées avec beaucoup de netteté, me paraissent très propres à ramener les esprits au degré de sang froid nécessaire pour considérer sous son véritable point de vue un objet important que l'on envisage depuis quelque temps avec un enthousiasme trop outré. à Paris, ce 20 Decembre 1761.' Baron de St Supplix, in *Le Consolateur*, 1763, spoke against 'agricultural fanaticism'. On the other hand, Fréron in 1764 though that 'cette agromanie, si on peut lui donner ce nom, ne peut produire que des avantages considérables.' On 'Agromania' see F. Brunot, *op. cit.* p. 197.

in his book[1] there are at least other shades of opinion. Of Tull and Duhamel he accepts certain important principles. He looks on artificial meadows with higher favour. But about the method of cultivation he does not yield.[2] Even the title of his work explains that his aim is to refute a certain method. He does it, by the way, very cleverly. He agrees that in France, lands do not produce a quarter of what they could. He attributes this evil to the 'fallacious routine of farmers, lack of meadows and cattle, inequality of taxes, etc.' Soil cultivation must, therefore, be improved, though not according to Tull's principles. French agriculture, regional methods of soil cultivation are 'founded on local practices of each canton'. These customs must as a whole, be respected. Therefore, says he, 'we have been particularly keen on combating M. Tull's method, as it directly destroys our local practice'.[3] There is in his attitude great respect for the 'ancient and noble' manner of cultivation, which the new system opposes only with bad results. La Salle's book is therefore that of an agricultural conservative, hostile in principle to too extreme novelties.[4]

We still meet in 1774 in a *Dissertation sur les Terres* statements which vigorously contradict Tull's notions, out of date at the time, on plant physiology, including the whole experimental activity followed for twenty years.

Our agronomers . . . will no doubt have noticed that they very wrongly attributed to the land what was often caused by the climate. They will, likely enough, have noticed the uselessness of their general

[1] *Manuel d'Agriculture pour le Laboureur, pour le Propriétaire et pour le Gouvernement . . . avec la réfutation de la nouvelle méthode de M. Thull*, par M. La Salle de l'Etang, Paris, 1764.

[2] As shown by an engraving by Cochin which illustrates the book and explained in *La Gazette Littéraire de l'Europe*, 1764: 'Une femme montre à un Laboureur un semoir à charrue pour l'engager à s'en servir; mais le génie de l'agriculture parait l'en détourner et lui conseille de suivre son ancienne méthode. On lit au bas ce mot de Caton "Ne change point de soc".' [3] See below, ch. v.

[4] He was vigorously answered by De La Marre, *Défense de Plusieurs ouvrages sur l'Agriculture*, Paris, 1765.

principles about agriculture. From these principles necessarily come a multitude of consequences, all opposed to established uses or accepted opinions. The sole instinct of the cultivator has accomplished more when he has followed the customs of the place where he lives, than when observing the instructions which may have been imported from outside. . . . I know how much it costs various farmers to have unadvisedly thrown themselves into the new speculations . . . they have learned how to distrust our agronomers' systems.[1]

These are the last reactions of a polemic which had had during twenty years a very real importance. Though it may still be read in the *Journal de Physique* that Dr Fabroni undertaking to 'demonstrate the evils of present agriculture' was 'an enemy, with some reason indeed up to a certain point, of multiple ploughings'.[2] The *Gazette Littéraire de l'Europe*[3] had already said that 'details of practice and theoretical methods' could be but of 'local use', often uncertain, always slow, and the controversy about the new cultivation was almost closed at the end of the century.

It is very often Tull personally or the 'English method' which has been criticized. As a matter of fact, as we have seen, the position of Duhamel originally was more moderate, and it is often his English model which they try to reach through him. Duhamel did not mean to break away entirely from traditional agriculture. Those who understood this moderate interpretation of the new system were also quite numerous at the time and, after all, their tradition was followed by the great school of agronomes of the nineteenth century. In an exhaustive work, posterior to the *Traité*, Dupuy-Demportes' *Gentilhomme cultivateur*, we find a sort of combination of the two opposed systems. This work has been considered, injudiciously as it seems, only as a 'wretched compilation'.[4] The various, too various, talents of its author[5] may

[1] 'Dissertation de M. Monnet sur les Terres', *Journal de Physique*, 1774, vol. IV, p. 183.
[2] 'Réflexions sur l'état actuel de l'Agriculture', *Ibid.* 1780.   [3] 1764.
[4] Article, 'Dupuy-Demportes', *Biographie universelle*.
[5] Journalist, novelist, poet and finally translator of English agricultural works.

justify the superficial judgment that agriculture was for him only one of his numerous hobbies. It is, however, immediately apparent that, as a compilation, it seems ready to welcome all the various currents of ideas of its time. In this respect, it is a precious source of information. On the other hand, the *Gentilhomme Cultivateur* was a book well received by its contemporary readers, often read and often quoted. In his *Année Littéraire,* Fréron quotes it with admiration and writes lengthy commentaries on each of its volumes.[1] Besides, its very form calls for a comparison with Duhamel's *Traité.* Like him, Dupuy-Demportes, the adapter of an English writer, Thomas Hale, is a follower of the English methods. He treats of the same problems as Duhamel (conscious perhaps of the interest awakened by the latter's work, and in an endeavour to win the same popularity for himself) but with more ambitious aims, as he studies rural economy, botany, political economy and so on. However, one can find in the *Gentilhomme Cultivateur,* a book claiming to have been acquainted with the methods used by peasants themselves,[2] interesting views about the new system.

From Tull he borrows unquestioningly the theory according to which plants feed on very small particles of earth. But he does not entirely follow his system of cultivation, which he finds too arbitrary. He maintains that peasant tradition may have some solid basis. We find, for instance, the very valid opinion that the fallow system must not be rejected altogether, as it gives the farmer 'time for a sufficient number of ploughings, for destroying bad weeds and for sufficiently preparing the soil to receive wheat'.[3] But for the preparation of a field, he is also a partisan of frequent

[1] Also in *Journal des Savants,* March 1762, pp. 225, 226; *Mercure de France,* 1762; *Jour. Oecon.* April, 1763, pp. 153-4.

[2] He gives precise instances of provincial techniques and corresponds with Comte de Maupeou and M. d'Epremesnil about artificial grasses. In *Année Littéraire,* 1763, vol. III, pp. 47, 48.

[3] His explanation of the fallow system can be compared with that of M. Lizerand, *op. cit.* p. 110.

ploughings and manurings. Plant cultivation with different ploughs he also advises. He goes even further and suggests, as Tull and Duhamel do, in connection with the seeding of lands, that seeds should be changed every year, and the drill used for sowing.[1]

Sometime later, Valmont de Bomare in articles in his Dictionary also expounds a moderate opinion.[2] He is, indeed, prejudiced in favour of England, 'la grande école de l'Agriculture', and thus gives a favourable view of Tull and Duhamel whose principles he approves of. He advises the use of the drill, 'which saves much seed by its manner of sowing, and so produces a better crop'.[3] But he is also a declared partisan of manures and gives a hint to Tull. 'A system of agriculture in which manures are not placed in the first degree of importance may be regarded as being dubious.'[4] If he praises artificial pastures, it is less because of their virtue in fattening cattle, than because they produce manure. 'These artificial meadows are regarded by all the best agriculturists as an essential and even unique means of improving our Agriculture.'[5] Tull advocated an agriculture adapted to large property, with a sudden and therefore costly, change. More moderate writers, on the other hand, suggest that methodical improvements without radical transformation will do as well.

The elaboration of a system in which tradition and new methods are combined was to triumph completely after 1770. We find the following in the *Journal de Physique* in connection with the debate on manures and ploughings. 'But manures would not be of any use for the earth without the benign influences of the atmosphere, that is to say, if the earth did not receive air, moisture and spirit (?) which are the essence of vegetation. This is the aim of the other side of cultivation, which consists of a multiplication of labours:

[1] *Gentilhomme Cultivateur*, vol. IV.
[2] *Dictionnaire d'Histoire Naturelle*, Paris, 1768.
[3] *Ibid.* Article: Blé, vol. I, p. 414.
[4] *Ibid.* Article: Fumiers, vol. II, p. 762.
[5] *Ibid.* Article: Prairies, vol. V, p. 190.

to turn the soil over, to burn it again, to divide it, to pulverise it. Without these operations, manures would be worthless.'[1]

Other factors also support the use of manures. From 1760 onwards, the question of cattle is of increasing importance, as is also the question of pasturing sheep, so important on account of their manure. On the other hand, people have a better knowledge of what is the actual practice in England, namely, a considerable use of manures. This may be the reason why we find at the end of the period a last thrust at excessive ploughing: 'In order to improve our lands, we multiply ploughings, till there occurs complete dissipation of the elements used by plant vegetation, and we destroy even the appearance of grass, which we call weeds.'[2] But he goes no further.

Considering the systems of Patullo, of the *Gentilhomme Culti-vateur*, of Fabroni, Rozier agrees that, after all, cultivation according to Duhamel's principles is the best. This was also to be the conclusion of Thouin and of the *Encyclopédie Méthodique*.

To sum up the whole debate, what remained of the English theory of soil cultivation in France at the end of the eighteenth century?

(1) Duhamel, offering the example of a careful and minute cultivation, tried to demonstrate that beautiful crops are obtained only through constant care and close attention. He criticized the exaggerated use of manures and insisted on the necessity of an excellent preparation of the ground. His views were accepted by agronomes of the following period and he thus ushered in the movement of agricultural improvements that was to take place in the nineteenth century.

(2) He tried to mechanize agriculture and make an applied

---

[1] 'Essai de météorologie appliquée à l'Agriculture' (Prix de la Société Royale des Sciences de Montpellier, en 1774), par l'Abbé Toaldo, *Journal de Physique*, 1777, vol. x, p. 253.

[2] Article: Amendements, *Cours complet d'Agriculture, théorique, pratique . . . par une Société d'Agriculteurs et rédigé par Rozier*, An. IX, 1801, vol. I, p. 194.

science of it, but almost completely failed in his own time. His teaching was heard only much later.

(3) Through his suggestion about new crops, copied from England, he was actually the first to initiate a movement for introducing into France plants hitherto not well-known or even totally new. This movement grew tremendously after him. Thus, thanks to his theory of making fallow land disappear by the use of artificial meadows, he may be called the father of rotative culture in France. At the end of the eighteenth century, it was only at its inception, but as a deep modification in traditional customs it was to continue developing and spreading ceaselessly.

Thus, in adapting Tull for France, he showed French farming the way that England was successfully following. The controversy about his book was fruitful. It led to the publication of numerous works, and forced the agronomic world, and perhaps also more farmers than is commonly realized, to experiment and think over the principles of agricultural science. The sensible eighteenth century, aware of its debt to the scientist, called him 'Benefactor of humanity'.[1]

[1] 'Review of the *Traité*', *Année Littéraire*, 1761, vol. II, p. 136.

PART THREE

AGRARIAN REPERCUSSIONS OF THE
NOUVEAU SYSTÈME

ELABORATION OF THE DOCTRINE OF ROTATIVE CULTIVATION AND ITS THEORETICAL CONSEQUENCES

THE mechanical aspect of the 'English Method' (i.e. actual manipulation of the land) was less important perhaps than the fact that it brought with it new ideas about crops already known and implied, with the introduction of new ones, a profound change in the agricultural landscape of the country.

It would certainly be going too far to say that English influence only was responsible for this movement. Indeed, other influences could also be felt.[1] All the same, though English influence was not exclusive, it was undoubtedly by far the most important.[2] And up to a certain point, as it is originally the source of the whole

---

[1] For instance, in Southern France, a very noticeable influence from Italy and Spain is marked in the works of agronomes of this area. See Reboul, *Discours sur les moyens d'encourager l'Agriculture en Provence*, Aix, 1770.

Although they cannot be compared with that of England, such influences as the following are worth mentioning: At the end of the century, Italy was very active in the field of agricultural research through the Academy of Georgophiles in Florence. It counted among its people famous agronomes like Fabroni and Toaldo. The Po valley was said to be one of the best cultivated parts of Europe. Switzerland also at the same time proved very active with the Société d'Agriculture of Bern, the *Mémoires* of which are full of useful indications, showing a wide range of interests. Sweden, too, had a developed agronomic school especially about questions of cattle raising. As for Germany and the Low Countries, we cannot speak of a positive agronomic school, at least being reflected in French works, but scattered information from these countries is often available.

[2] Often it is only a vehicle for methods hitherto ignored in France. For instance, rotation of crops, although it had its origin in Flanders, was introduced into France through England (see Chaptal, *De L'Industrie française*, Paris, 1819, p. 142).

economic movement in the second half of the century, so, one can trace it all through the movement of agrarian reforms that occurred in connection with the new economic trends of the time. That is why it may be not without interest to endeavour to discern among these the part played by English-inspired tendencies.[1]

The new system opened two ways to agriculture. Once the solution of breaking the land either with ploughing or with manures was agreed, the wider consequence of Duhamel's theory very soon appeared. The mechanical principle of it theoretically involved the constant occupation of the land by cultivated crops (in the case of Duhamel, it was wheat), which itself in suppressing the fallow made necessary the existence of extensive pastures of artificial grasses. Whatever the intrinsic value of Tull's theory, the fact is that its translation raised this problem, for the first time, for the consideration of French agronomers. The relations which were soon found to exist between occupation of the soil by corn on the one hand, and the new crops on the other (for instance, alternation of wheat and turnips or sainfoin) had been, it is true, suspected before Duhamel. Liger had noticed that certain plants could be of value in improving the yield of the succeeding crops.[2] The idea of an uninterrupted cultivation was not altogether unknown.[3] But such an agriculture had neither a definite method nor a clearly realized purpose. Often the peasant tried to get as much as possible out of his land, especially when he was established on a short-term lease, and therefore did not care much about

[1] G. Lefebvre (*A.H.E.S.* 1929, p. 519), after Wolters, partly attributes to the English influence the agrarian transformations inaugurated by the French Government.

[2] Liger, *Economie générale de la Campagne ou nouvelle maison rustique*, Amsterdam, 1701, vol. I, p. 253.

[3] In the North of France (G. Lefebvre, *Les Paysans du Nord pendant la Révolution Française*, Lille, 1924) and in the South-West (Duhamel, *Eléments d'Agriculture*) the practice could be found of cultivating some crop during the fallow year ('dérober la jachère'). Chaptal, *De l'Industrie française*, Paris, 1819, vol. I, p. 142, shows the migration of the system of rotation from Flanders to England, thence to France.

the consequences of his method of cultivation.[1] In other parts of the country, agricultural experiments were governed by an almost blind empiricism. Nothing was actually founded upon a scientifically formulated theory. Up to 1750 all researches were generally directed towards an improvement in the fallow-field system, not towards its absolute suppression, whereas at the same time its agricultural evolution was leading England from the old system to a scientific rotation of crops.

Duhamel, in fact, opens the debate about rotative cultivation. He was, in this respect, supported by Forbonnais. But the Norfolk letter in the *Eléments du Commerce* was limited to a description of an agriculture which did not allow the land to lie fallow. It was in Tull's *Treatise* that the French agriculturists became conscious of an almost perpetual cultivation by means of a perpetual reconstitution of the soil. After him, Duhamel seems to follow the same track. The part of his work in which he tries to explain the improvement of lands with certain crops (sainfoin, for instance) is very obscure and inaccurate.[2] However, he states: 'Let us then conclude that it is possible to obtain a good crop of wheat in the same land, every year. It is only necessary to repeat the process of ploughing.'[3] These somewhat bold views were, however, tempered by a great number of very accurate observations.

Although it seems from the *Traité* that Duhamel did not exactly understand the rotation of crops, nevertheless he advises the introduction of artificial fodder into the traditional order of the three-field system. We have seen him devote chapters to

---

[1] 'Il résulte d'une si courte limitation dans la durée des baux, que les changements de fermiers s'opèrent très fréquemment, que celui qui finit un bail, s'il n'en obtient un nouveau de bonne heure, ne cultive qu'imparfaitement les trois dernières années de peur d'être augmenté en raison des améliorations qu'il aurait faites', quoted in Calonne, *La vie agricole sous l'Ancien Régime*, Paris, 1883—The argument is still met with today. See A. Garrigon-Lagrange, *Production agricole et économie rurale*, Paris, 1939, p. 38.

[2] He seems, even, not to believe in alternation of crops (*Traité*, vol. I, Préface, p. xxii). But he amended his view later.   [3] *Traité*, vol. I, ch. IV, p. 46.

sainfoin, lucerne and roots. There is a sort of primitive idea of rotation in the description he gives of an alternation between turnips and corn.[1] He certainly advocates a change in the traditional system, as he is obliged to answer the objection that if all the land is constantly occupied by crops, it will then be impossible to breed cattle: 'Je crois que cette objection n'a point échappé à M. Tull et qu'il compte destiner une partie de ses terres à élever des raves et des navets et d'autres herbages que les Anglais appellent des pâturages artificiels, qui, suivant lui, subviennent au besoin du bétail.'[2] The history of Norfolk, amply demonstrates the excellence of these new crops.

> Entendue considérable de terre, regardée auparavant comme mauvaise et stérile, est à présent devenue une des plus fertiles et des plus abondantes de l'Angleterre. Le principal moyen par lequel on a opéré un changement aussi étonnant est fondé sur des principes assez analogues aux nôtres, surtout pour ce qui regarde les pâturages artificiels qui ont fourni le moyen de varier fréquemment la culture. A la vérité la même enclos n'est jamais ensemencé de blés pendant plusieurs années de suite; mais après une ou deux récoltes de ce grain, on fait rapporter à la terre des herbes propres aux pâturages, principalement des turnips ou gros navets. . . .[3]

Such is the importance in this respect of Tull's book in France. It clearly shows the advantages of the new crops and the place they must have in the traditional order of the breaks. But Duhamel was not the only one to praise the new crops. Others amplified what he had said and brought the question to a high degree of accuracy.

The first book which actually treated of rotative cultivation as such is Henry Patullo's *Essai sur l'Amélioration des Terres*. Patullo had, no doubt, the closest links with the Physiocrats, as his book is dedicated to Mme de Pompadour, patroness of the

---

[1] This is indeed the first example of a reasoned use of the different crops. Actually such practices were sometimes known in certain districts. Duhamel did not only widen their range but also based their application on the new principles.

[2] *Traité*, Préface, vol. x, pp. xvi ff.   [3] *Traité*, Préface, vol. iv, p. xxi.

School,[1] and he continues, in a way, the work of Duhamel whose theories he advocates. He is a remarkable figure in the French agronomic world, as he is undoubtedly English by birth. The Abbé de Petity gives some valuable information on this subject:

M. Patullo, par la reconnaissance de l'asyle qu'il a trouvé en France depuis plus de dix années, et des bienfaits du Roi dont il jouit, a voulu témoigner par cet écrit, le désir qu'il a de contribuer à la perfection de notre Agriculture en nous faisant part des moyens qui font fleurir cette branche importante de l'Economie Politique, en Ecosse, dont il est originaire, et en Angleterre où il a passé la plus grande partie de sa vie.[2]

In this way, English influence penetrated directly into the field of agricultural research in France.[3] Duhamel's 'English' agriculture and Patullo's work are closely connected. Duhamel's principles, and his relations with English thought, naturally led him to the theory of rotation. But Patullo went a step further, as he definitely included artificial fodder within the traditional order.

This order, for instance, near Bayeux, in a country of rich farm-land, consisted in the alternation of wheat and spring crops over a period of five years. The last year included a crop of oats and one of clover. After this, the earth became a natural pasture for an undetermined length of time. Patullo changes this traditional method into another which combines the cultivation of grain and artificial fodder: (1) wheat and burning of stubble; (2) turnips and one ploughing; (3) white peas and one ploughing; (4) turnips

---

[1] 'La protection décidée que vous accordez à ceux qui s'appliquent à l'étude de la Science œconomique...', Dedicatory epistle to Mme de Pompadour. (Dupont de Nemours, *De l'exportation et de l'importation des grains*, 1764.)

[2] 'Bibliographie agronomique', *Encyclopédie élémentaire*, p. 615. In fact a paraphrase of Patullo's Préface in the *Essai*. Petity uses this notice, it seems, as a kind of advertisement.

[3] Mainly through a definite influence exerted by Scotch exiles in France. We find at the same time a certain 'milord Ogilvy', whose husbandry writings were known and appreciated. In his *Mémoire sur les Semoirs*, 1761, he supports the new cultivation and Abbé Soumille's new drill. Also a more famous name, Holker, who found refuge in France in 1748, 'capitaine en second au régiment d'infanterie d'Ogilvy', a friend of the Trudaine and who introduced into France English industrial processes.

and two ploughings; (5) barley and one ploughing; (6) clover and one ploughing; (7) barley and two ploughings; (8) wheat.

This combination is one of several which may be found in the *Essai sur l'Amélioration des Terres*.[1] Its influence was to be very great and Patullo was to be quoted throughout the rest of the century as one of the famous agriculturists of his time, and the founder of a definite agricultural system.[2] He endeavours in fact to transfer English farming to France. Besides expounding the theory of rotation, the *Essai* is a handbook of English farming for the use of the French.[3]

The author fully realized the audacity of his undertaking, but was also fully convinced of its success. In the exposition of his doctrine, we may see how much the English system was intended to change the agrarian state of the country.

Je m'attends bien qu'on pourra penser que c'est trop astreindre l'Agriculture à un système particulier que de vouloir enclore les terres de toute espèce, et en semer toujours une certaine quantité en herbages artificiels, une autre en orge et une autre en froment deux années de suite, excluant les seigles et surtout les avoines employées partout à la nourriture des chevaux, ne faisant mention d'aucune pâtures, quoiqu'il soit difficile à croire que les bestiaux de toute espèce puissent être toute l'année, nourris sainement à l'étable, et qu'on ne donne point la méthode d'y employer les fourages artificiels; ne se servant que des chevaux aux travaux dont on donne le plan et semblant en exclure les bœufs.[4]

---

[1] Reproduced in Duhamel's *Eléments* and also in Article: Culture, *Encyclopédie Méthodique* (Art aratoire et jardinage), p. 71.

[2] We find the following mention in the *Journal de Physique* (1773),vol. I, p. 243, about Patullo's *Essai*: 'La réputation des ouvrages que ce grand agriculteur a publiés augmente nos espérances et nous fait désirer de voir cet ouvrage écrit en Anglais, traduit dans presque toutes les langues de l'Europe. Il sera sûrement accueilli avec autant d'ardeur que l'ont été les autres traités de cet Agronome.' The review alludes here to Patullo's second work, *An Essay upon the Cultivation of the Lands and Improvement of the Revenues of Bengal*, by Henry Patullo, London, 1772.

[3] Patullo actually intends to 'décrire aussi exactement que la brièveté . . . le pourra permettre, la pratique d'amélioration et de culture moderne en Angleterre', *op. cit.* pp. 1-4 and 9-10.　　　　　　　　　　　　　　[4] *Ibid.* pp. 122-4.

The means to be employed were: (1) improvement of all lands by varying them and an appropriate use of the known manures; (2) enclosure of all the fields and division of the territory of a farm into closed and separate enclosures; (3) use of one-half or one-third of the land for artificial fodder; (4) alternative succession of cultivation from grass to plough and from plough to grass, an order which preserves fertility; (5) feeding a greater number of animals, and entire consumption of fodder in order to obtain fertilizers.[1]

Duhamel's mechanical theory is not applied in its entirety in this agricultural programme. But as it certainly inaugurated a new conception in farming many points were tested, discussed and frequently preserved, thus introducing some change into the traditional system.

At the same time accounts of experiments were published in which a concern to use artificial grasses methodically may be found. An article entirely devoted to this question enables us to understand the stages through which crop rotation passed. Its author, though he acknowledges the worth of Duhamel's agriculture, sets out to provide 'a system of farming the worst districts without, however, incurring too great expenses on oneself'.[2] He takes as an example Champagne, where 'the lands are divided, as in other countries, into three breaks'.

His belief in the introduction of artificial meadows into the three-field system, has led him, as he explains, to make an agrarian change in his own estate, so that henceforward it is divided into four parts, each one successively occupied over a period of five years by sainfoin. Then, after twenty years, the artificial meadow will have passed through all the parts of the estate.

This is an over-simplification of the new theory and its too rigid application. It may even never have been actually attempted. Yet it is a sign that the new ideas were spreading, as such

[1] *Essai*, pp. 122-4.
[2] 'Prairies artificielles' (Review of La Salle de l'Etang's book), *Nouvelliste Oeconomique*, 1756, vol. XVII, p. 75.

experiments were repeated and information on the improvements brought about by the use of artificial meadows was regularly published. Several books of merit appeared, all derived either from Duhamel or Patullo, on whom they enlarged or whose suggestions they completed. The most famous are those by Delisle,[1] La Salle de l'Etang, Gilbert[2] and Cretté de Palluel.[3] The clarity of their presentation and the actual value of their content make them truly important works. In spite of their repeated exhortations, however, resistance and prejudice were still met with. Some purposely exaggerated the intentions of the agriculturists. The Baron de St Supplix, for instance, complains that 'Among this crowd of economic writers who speak so much about the flourishing state of England, very few are acquainted with it,' and later states, 'It so happens that in Norfolk there are artificial meadows almost everywhere. And it is the same over a large area of our own Champagne and certain other districts; but indiscriminate sowing of artificial meadows would be as great a folly as that of the man who wanted all the French towns to be sea-harbours.'[4]

However, the importance of artificial feeding stuffs became increasingly apparent as the century went on. A good agronome could no longer ignore them. In his report on English farming[5] Roland de la Platière points out that turnips were sown in a 'break which would otherwise lie fallow: it improves it for the crops of the following year'. Until the Revolution, such remarks were common. Without doubt, the theory of rotation was hard to accept completely as it opposed long-established traditions and meant a profound change in French agriculture. The system also presented itself as the result of scientific experiments, and had

[1] *Mémoire sur le Ray Grass et le Red Clover*, 1761.

[2] *Traité des prairies artificielles*, Paris, 1790.

[3] *Traité des prairies artificielles*, Paris, 1801.

[4] 'Le Consolateur: pour répondre à la théorie de l'Impot', review in *Année Littéraire*, 1763, vol. IV, pp. 155-6.

[5] *Journal de Physique*, 1779, vol. XIV, pp. 51-100.

therefore to be carefully expounded. Thus most of the second-rate books were but compilations of the best ones and often repeated one another.[1] But the strong resistance to any change in the agrarian distribution of the land (involved in the new system) was the cause of a continuous output of books by specialists and also of research in the improvement of methods. In 1778 the 'Société Libre d'Emulation' was setting forth the problem very clearly in asking:

Quels sont les moyens les plus avantageux pour rendre plus utile et plus fructifiante l'année de repos que plusieurs agriculteurs croient devoir donner aux terres à grains sous le nom de jachères ou de guérêts? . . . Les concurrents observeront qu'il existe encore un grand nombre de propriétaires ou de fermiers persuadés que les années de repos alternatif sont utiles ou même absolument nécessaires aux terres à grains. Mais que d'autre part, en Angleterre, en Hollande, en Suisse et dans plusieurs Provinces de France, on a déjà supprimé les années de repos ou de jachères, par la méthode nouvelle de varier les assolements, qui consiste à cultiver alternativement dans le même champ, des grains, des légumes, des fourrages et des racines; on assure que la terre, bien loin d'être épuisée par ces productions continuelles et variées, n'en devient au contraire que meilleure et plus fertile.[2]

Later, at the very end of the century, arguments in favour of crop rotation were multiplied by allusions to English successes in this field. In the famous agricultural circles of the time, the importance of the English method was often recalled by those who had visited England.[3] Baert, about ten years later, described with admiration the state of Norfolk, 'still nowadays the best cultivated

---

[1] For instance: Thierriat, *Instructions familières en forme d'entretiens sur les principaux objets qui concernent la culture des terres. . . .* Paris, 1764; and *L'Abondance rétablie ou moyens de prevenir en France la disette des bestiaux en même temps qu'on rétablit la fertilité de la terre*, Paris, 1769. Such pseudo-scientific books or pamphlets are innumerable.   [2] *Journal de Physique*, 1788, vol. ii, p. 292.

[3] Dumont de Courset expresses a very moderate view when he writes, 'Ce n'est point aussi la division des terres en soles que je trouve mauvaise; elle est même nécessaire vu la disposition des biens de ce pays, le nombre de ses habitants, la quantité insuffisante de ses bestiaux et la modicité de ses engrais; mais c'est l'exactitude scrupuleuse dans la conservation de ce partage qui fait une espèce de loi, que je voudrais dissiper ou au moins modifier', *Mémoire sur l'agriculture du Boulonnais*, Boulogne, 1784, p. 153.

of the English counties, where one can most profitably study the principles of an extremely difficult art, the only solid basis of which is a long series of observations'.[1] And although the *Nouveau Cours d'Agriculture* at the opening of the nineteenth century still spoke of rotation as an 'art so useful and so little known', the prolific articles of Rozier, of Tessier and of Sonini de Manoncourt contained in an orthodox form the henceforth classical theory of rotation and condensed the results of fifty years of discussion and experiment on this method imported from England.[2]

This advocated alteration of land distribution and land cultivation meant in fact a change in French rural life as it existed. The new system struggled not only against the traditional partition of the land as being inadequate for carrying out the new method of farming successfully, but also against all sorts of easements and interpenetrations which bore on the relations between two pieces of land or two districts. Marc Bloch has drawn a masterly picture of these rural problems inherited from feudal times, which maintained a vigorous opposition to theories reaching a solution as far as agriculture is concerned, but disregarding the inevitable disorganization within a rural society which was built upon a community basis and common interests.[3]

Two main factors helped to retain a low agricultural standard in most parts of the country: the right of pasture and the existence of commons..

The history of these two traditional phenomena of French rural life need not be told again.[4] It may however, be interesting

---

[1] *Tableau de la Gde. Bretagne, de l'Irlande et des possessions anglaises*, Paris, An. 8, vol. III, pp. 242-3. The book published in 1800 was written in America shortly after 1792, and summed up observations made before 1789.

[2] See a good account of the question in De Pradt, *op. cit.* vol. I, pp. 139 ff.

[3] The best study on the question is Marc Bloch, *Les Caractères originaux* Oslo, 1931.

[4] M. Bloch, 'La Lutte pour l'individualisme agraire dans la France du XVIIIe siècle', *A.H.E.S.* 1930, pp. 328 ff. For other studies on the same question, see Bibliography.

to find out to what extent their abolition was contemplated after the propaganda in favour of the new husbandry and also in direct imitation of England.

Here again England was to be the inspiring and concrete example.[1] Indeed, certain very limited areas in France had already begun a tentative movement of enclosures, but it was considered as a provincial peculiarity rather than an example to be followed.[2] England, on the contrary, was the country where the combination of enclosures and the new husbandry was most striking. Although Duhamel did not say much about it, Dangeul, at about the same time, wrote that in England, 'barren or badly cultivated commons, sterile or desert-like pastures, became, thanks to their enclosure and separation by hedges, fertile fields and rich pastures . . . universal experience teaches that lands so cultivated have doubled their productive capacity'.[3] It was in Norfolk mostly that the French agriculturists looked for examples. Patullo mentions the fact and vigorously upholds the necessity of enclosures. 'La pratique d'enclore les terres a commencé depuis longtemps en Angleterre et y est maintenant presque générale. On a éprouvé que ce seul avantage ne manque guère de doubler la valeur du fond.'[4] Desplace's answer shows that the value of enclosures was still not clearly realized: 'L'auteur de l'Essai sur l'Amélioration des Terres, prévenu peut-être en faveur des usages de sa patrie croit

[1] President Musac of the Parliament of Metz stated in 1762 at the Metz Academy of Sciences, 'Nous ne douterions peut-être pas nous-mêmes que cet usage est mal si les Anglais ne nous avaient appris à connaître les forces de la terre . . . sans l'ombre que les tableaux de leur agriculture ont jetés sur la nôtre, nous pourrions croire encore avec nos Pères, que ces droits de vaine pâture ne forment pas une servitude.' (M. Bloch, *op. cit.* p. 354.)

[2] Inasmuch as there was no connection between these traditional enclosures and a modernized system of husbandry (see John Clapham, *The economic development of France and Germany, 1815-1914,* Cambridge, 1936, pp. 6, 7, and a typical remark of Blaikie (*Diary,* Paris, 1931, p. 145).

[3] *Remarques sur les avantages et les désavantages de la France et de la Grande Bretagne. . . . Traduites de l'Anglais du Chevalier John Nickolls,* Amsterdam, 1754.

[4] 'De la Clôture des Terres', *Essai,* p. 30.

ne devoir trop recommander cette clôture des tous les champs à laquelle nous ignorons ce qui peut avoir donné lieu en Angleterre.'[1] Dupuy-Demportes again takes up the question and writes: 'Tout cultivateur qui voudra amender les plus mauvais sols, s'il commence par la clôture, sortira avec succès de son entreprise. Les Anglais, par ce moyen, ont su mettre à profit les sables mouvants de la Province de Suffolk' and he adds: 'si l'usage des clôtures s'établissait en France, comme il existe déjà dans une partie de la Normandie, non seulement nous aurions des foins et des pâturages bien supérieurs à ceux que nous donnent les prés ouverts, mais nous pourrions encore cultiver plusieurs espèces de prairies artificielles sujettes à différents inconvenients dans les prés ouverts.'[2]

The connection between the proposed establishment of enclosures and the new agricultural conceptions appears in some of the reactions against Bertin's Edicts.[3] The minister was convinced, for instance, that the right of pasture was responsible in Lorraine for the existence of the three-field system, condemned by the agronomes. The Conseil Souverain of Lorraine, on the contrary (with the exception of La Galaizière, a disciple of the new ideas), where agricultural conservatism prevailed, opposed the Edicts on the grounds that the right of pasture was a necessary element of the traditional three-field system.[4] No agreement having been reached on the basic principles of agriculture, these shades of opinions are understandable. There were undoubtedly other grounds for opposition besides the mere preservation of selfish interests. This opposition to enclosures was due sometimes to the fact that the results of the new husbandry were variously assessed by people inhabiting the same region. M. de la Case, Premier President in

---

[1] Desplaces, *op. cit.* pp. 185, 186.

[2] *Gentilhomme Cultivateur*, p. 24.

[3] Marc Bloch, *Les Caractères originaux, op. cit.* pp. 225-35. See Duhamel, *Eléments*, vol. II, p. 377: 'La vaine pâture et le parcours forment un obstacle invincible aux progrès de l'agriculture.'

[4] H. Sée, *La France Economique et Sociale*, Paris, 1925, p. 37.

the Parlement of Pau, opposed the Edicts on the grounds that artificial grasses could not grow in Bearn, whereas at the same time, the Intendant d'Aisne said exactly the contrary.[1]

Yet the new agricultural principles seemed too firmly established, and the example of England too conclusive, to abandon the theory of enclosures. Enclosing, hitherto a means of better cultivation of the soil, was rapidly to become, so to speak, an end in itself. An enclosed country was synonymous with a rich agriculture.[2] 'Dans les pays où l'Agriculture et l'Industrie brillent de toutes parts, l'on ne voit que des clôtures; tout le canton ressemble à un échiquier.'[3] Various writings on the technique of enclosing, on hedges, were multiplied and most of the examples given were borrowed from English sources. The economists, also, linked these agricultural conceptions with politics. The political state of England was associated in their minds with that agricultural individuality. Abeille saw in the agrarian transformation another aspect of the political revolutions which 'avaient rendu le peuple Anglais libre'.[4]

The failure to achieve, in fact, the agrarian change by means of enclosures did not appear to its contemporaries as striking as we consider it today, with our historical perspective. At the end of the period, as knowledge progressed, the question of enclosures reached a climax. In 1786 the Duc de Liancourt still spoke of English agriculture as 'incomparablement plus florissante que celle de France', and as being too little known; and he emphasized the

[1] *Ibid.* p. 36. Bomare (Article: Prairie, *Dictionnaire d'Histoire Naturelle*) shows, although vaguely, the conflict between the state of things and the theory: 'Quoique un très grand nombre de personnes convienne de la supériorité des prairies artificielles, il y en a cependant beaucoup qui ne peuvent se résoudre à leur sacrifier les pâtures, c'est à dire, les terrains qui sont en jachère.'

[2] See M. W. Gilpin, M.A., *Voyages en différentes parties de l'Angleterre*, traduit de l'Anglais par M. Guidon de Berchère, Paris-Londres, 1789, vol. I.

[3] *Mémoire sur l'Agriculture du Boulonnais et des cantons maritimes voisins par M.d.C.*, Boulogne, 1784, p. 187.

[4] Quoted by Bloch, *La Lutte pour l'individualisme*, p. 354, from *Mémoire présenté par la Société Royale d'Agriculture*, 1789, p. 32.

need of enclosures.[1] Lazowski, some time later, devoted lengthy passages to the subject and stated that in England the right of pasture had been abolished by the wish of two-thirds of the *propriétaires* (landowners).[2] Tessier summed up all these observations in his exhaustive article, *Clôture*, which said: 'Aucune Nation n'a porté l'art de faire les clôtures à un si grand degré de perfection que la Nation Anglaise ... c'est à leur établissement que l'on peut, en grande partie attribuer l'état florissant de leur Agriculture.' According to him, the advantages of enclosures in France would prove to be the same as those noticed by French travellers in England.[3] He expresses the same views as the theoretician of hedges, Amoreux, who writes, 'Les parcours sont un obstacle à l'amélioration de champs, à leur plus grand produit, à leurs récoltes variées. Disons plus, les parcours sont la ruine de l'Agriculture.'[4] Thus, if the right of common pastures 'de tous les droits le plus abusif et qui s'oppose le plus aux progrès de l'agriculture'[5] is abolished, better cultivation will result under the protection of enclosures.

Almost the same argument was used about commons. The 'Gentilhomme Cultivateur', Desplaces himself, indicated that 'peut-être les progrès de l'agriculture en Angleterre dont nous sommes si jaloux ne viennent que de ce que les Communes ont été partagées'.[6] There is a similar remark in the *Encyclopédie Méthodique* which states, 'On assure que l'Angleterre date les brillants succès de son agriculture de l'époque du partage des

---

[1] Quoted by Bloch, *La Lutte pour l'individualisme*, p. 233.

[2] Pigeonneau et de Foville, *op. cit.* p. 207.

[3] In 1804 Parmentier, back from his visit to England, stated in the Société d'Agriculture, 'J'étais partisan des clôtures avant d'avoir voyagé en Angleterre, et, depuis mon retour, je suis convaincu plus que jamais, qu'elles sont une des causes auxquelles ce pays doit l'état florissant de son agriculture', in *Mémoire d'Agriculture*, vol. VI, an. XII, 'Mémoire sur les Clôtures', p. 301. It is to be noticed, however, that Baert found enclosures in England, actually less extended than was commonly thought.   [4] *Traité sur les haies*, p. 210.

[5] De Fresne, *Traité d'Agriculture*, Paris, 1788, vol. I, p. 93.

[6] Desplaces, *op. cit.* pp. 175, 176.

communes et de l'abolition du parcours.'[1] This problem, one of the major undertakings of the *ancien régime*, was closely connected with the introduction of the English system.

The zealous support of the 'Seigneurs', generally, of the partition of commons has led to indictments of rapacity from modern historians.[2] Although there is much truth in it, it must not be forgotten that partition of commons and reclamation of waste lands were a natural consequence of the new agricultural principles, which, in involving a more intensive production of fodder and claiming to permit the cultivation of the worst soils, theoretically showed the existence of commons and waste lands to be unnecessary and pernicious.[3] Besides, imitation of England on that score was evident.

The agronomes, therefore, had founded in theory what the *seigneurs* later tried to put into practice.[4] The whole movement was initiated by them. The idea was fully developed by Tessier: 'Mr l'Abbé Rozier observe avec raison que nos bonnes terres actuelles ressemblaient autrefois a des Communaux. La culture les a rendues fertiles.'[5]

With the exception of those lands absolutely incapable of cultivation, partition and clearing must then be carried out. 'Les partisans des défrichements allèguent en leur faveur l'état florissant de l'Agriculture Anglaise. Il disent qu'elle a changé de face depuis que les Cultivateurs de cette Ile ont employé à défricher les landes, les Communes, les mauvaise pâtures, la gratification accordée à l'exportation des grains.'[6]

---

[1] Article: Communes, vol. XXI, p. 383.

[2] J. Jaurès, *Histoire socialiste de la Révolution Française*, Paris, 1922, vol. I, p. 234, after him, Lefebvre, Bloch.

[3] See H. Sée, 'Le partage des biens communaux à la fin de l'Ancien Régime'. *Nouvelle Revue Historique du droit*, 4e série, 1923, vol. II, p. 49.

[4] On that point Jaurès (*op. cit.* p. 234) quotes Merlin de Douai speaking of the seignorial action as 'le masque d'un faux zèle pour le progrès agricole'.

[5] *Encyclopédie Méthodique*, p. 381.

[6] *Ibid.* Article: Défrichements, vol. IV, p. 38.

This idea of the restoration of lands to cultivation had begun to be commonly accepted since Duhamel's book. The design of the four-coultered plough and of several other ploughs for breaking up the lands, deriving from it or fulfilling the same purpose, shows that the new system was meant to lead to a greatly increased cultivation of all kinds of land.[1]

Propaganda for reclaiming barren lands or commons was supported by the arguments for the new method. The *Journal Oeconomique* (1751) devotes an article to land clearance in Norfolk and the method used in effecting it. It was then that Matthew Yelverton of Portland became famous for having obtained a superb crop from previously barren land and so won the prize of the Society of Dublin.[2] The Marquis de Turbilly's famous book[3] minutely described the experiments performed by this Townshend of Anjou, some of which had a definite English flavour, and which were to have a very great influence on government circles. A great enquiry about the partition of common land began in the same year as that about the abolition of the right of pasture, in 1766.[4] In 1761 a Royal Edict encouraged and granted privileges to those who would undertake to break up and reclaim barren lands.[5] Such were the material consequences of the propaganda for the new system.

In 1770 Comte d'Essuille's *Traité des Communes* summed up the results of cultivation in some of the partitioned commons. It showed an increase of population, livestock and produce. Thus,

[1] One of the introductory chapters of Duhamel is 'Du Défrichement des Terres', vol. I, ch. VII, pp. 66 ff.

[2] *Jour. Oecon.* 1752; Duhamel, *Traité*, vol. VI, ch. I; Yvart, in his report on England (*Mémoires d'Agriculture*, vol. X), commenting on this famous and popular symbol of the 'défrichements' in France in the eighteenth century, writes about the miraculous wheat crop, 'il a été malheureusement reconnu qu'elle n'avait existé que sur le papier', p. 84, n. I.

[3] *Mémoire sur les Défrichements*, Paris, 1760.

[4] M. Bloch, *loc. cit. A.H.E.S.* pp. 355 ff.

[5] See H. Sée, 'La mise en valeur des terres incultes, défrichements et desséchements à la fin de l'Ancien Régime', *R.H.E.S.* 1923, vol. XI, pp. 62-81.

the arguments which hinted that a decrease of crops and cattle would follow partition of the commons were finally refuted. The Abbé Carlier also pointed out that 'l'opinion qu'ils ont occasionné une réduction du nombre des troupeaux et, par conséquent, du nombre des individus qui les composent, est dénuée de fondement'.[1] This was also the opinion of Baert, who wrote at the end of the century about Norfolk:

des terrains enclos et saignés avec intelligence sont bien préférables pour les bestiaux, à de mauvais marais remplis d'eau croupissante, que personne ne se donne jamais la peine de faire écouler et où ils périssent de pourriture; et celles qu'on cultive en grains sont périodiquement converties en prairies artificielles, dont le produit joint aux pailles des récoltes précédentes repousse toute comparaison avec celui de mauvaises landes presqu'entièrement couvertes de fougères ou de joncs marins.[2]

[1] Mémoire, *Journal de Physique*, 1784, vol. XXIV, p. 107.
[2] Baert, *op. cit.* vol. III, p. 254.

CHAPTER VII

SOCIAL, JURIDICAL AND POLITICAL
IMPLICATIONS OF THE *NOUVEAU SYSTÈME*

THE agronomes did not only wish for the establishment of their agricultural technique. Although they were mostly concerned with problems of pure practice, they often ventured to submit solutions for its acceptance in rural society. Their demonstrations were more rudimentary than those of the representatives of the physiocratic 'school'. Yet they were interesting in so far as they were intimately connected with problems of pure technique; they often possessed a sort of earthy flavour and frequently looked to England for guidance and confirmation.

They knew the system of taxation, as it existed, to be a major obstacle to the acceptance of a new method, which involved great expense. Thus, parallel to the attacks led by people like Le Trosne[1] and Boncerf,[2] against feudalism and feudal rights, are the agriculturists' complaints against the excessive taxation imposed on the peasants. But, in contrast to the gloomy colour of rural life as it is presented in works of general scope, quite a different picture can sometimes be obtained from local information. The Abbé Roger Schabol gives an interesting picture of the Montreuil territory.

Ailleurs, ce sont de vastes oseraies, des prés, des luzernes, des sainfoins, des blés, du seigle pour faire du pleyon, des grains de toute nature. On trouve de côté et d'autre, à l'écart, de petits cantons de terre ménagés qui forment des bouts de pépinières. . . . Là, me dit-on, nul

---

[1] *De l'Administration provinciale*, Basle, 1788.
[2] *Les Inconvénients des droits féodaux*, Londres, 1776.

n'est oisif ni exempt de travail. . . . Aussi sait-on que le Village de Montreuil paye jusqu'à quatre-vingt mille livres de taille.[1]

D'Epremesnil[2] similarly pictured the Pays de Caux, 'où il n'y a pas un pouce de terre qui ne soit cultivé avec le plus grand soin. . . . Quelle prodigieuse quantité de Fermes! Parcourons ces fermes; nous verrons les Fermiers, la plupart riches, tous aisés.'[3]

It must be observed that these examples of thriving districts, however, showed two different aspects of agricultural structure. In the former, agriculture was somewhat like gardening and belonged to small-scale farming.[4] The latter was an enclosed country in process of modernization. But the important fact is that they had both reached the stage of a complex agriculture where artificial grasses were combined with corn production. This seemed to prove that the new doctrine was based on principles which, in a district, attracted wealth and activity in spite of the taxes. Indeed, though these are examples of two particularly favoured districts, we can find here and there instances of others, under transformation and gradually improving, which followed the new system. In certain parts of Lorraine, for instance, the inhabitants had gradually developed from the community stage to one of individualism. They were applying a method rather similar to the 'new system' 'qui depuis quinze ans a diminué la misère du comté de Bitche'. In 1783 the Conseiller Boutier thus summed up the situation: The inhabitants 'ont unanimement

---

[1] 'Sur les Villages de Montreuil, Bagnolet, Vincennes, Charonne et autres villages . . . au sujet de la culture des végétaux par l'Abbé Roger', *Nouvelliste Oeconomique*, 1755, vol. VII, p. 21. The Abbé Roger Schabol was the principal name of French gardening in the eighteenth century and the exponent of the Montreuil practice.

[2] Father of the enthusiastic *conseiller* in the Paris Parliament at the beginning of the Revolution.

[3] 'Correspondance sur une question politique d'Agriculture', Amsterdam, 1763, *Année Littéraire*, 1763, vol. VII, p. 107.

[4] Besides, Montreuil was in the neighbourhood of Paris and the influence of such a great centre of consumption explains partly this mode of intensive cultivation.

demandé d'être maintenus dans leurs usages champêtres, la liberté de semer des prairies artificielles sans les clores, de jouir du regain et de la seconde herbe sur les prés, de greffer les arbres champêtres au profit des propriétaires'.[1]

In most parts of the country, however, fear of increased taxes made the peasants reluctant to change their accustomed habits. Not only might the *taille* be increased, but also the introduction of artificial grasses raised the question of *dimes novales* (new tithes).[2] English laws and customs were therefore cited by the agronomes to criticize the French system: 'Le paysan anglais paye de fortes impositions à l'Etat mais il ne dépend pas d'un intendant, d'un subdélégué, d'un commis, de l'augmenter d'un denier. S'il en était de même en France, le Royaume ne serait pas tombé dans le malheur et le dépérissement.'[3] At the end of the century, Baert still considered that the taxes paid by the English farmers were very heavy.[4] But Tocqueville showed well the difference between England and France on that score, and showed why the English peasant could put up with those taxes.[5] Roland, comparing the natural pastures of England and France, writes, 'J'ai remarqué avec étonnement que dans la plupart de nos provinces, tous les Prés ont des siècles. On les fume, on les cendre, on les arrose'[6] but they are never ploughed, for fear of an increased tithe. And he continues to criticize the French mode of imposing taxes:'Ainsi on ne cultive point ici, dans la crainte d'un impôt qui n'est pas seulement mis sur le produit, mais sur le travail, mais sur les semences, mais sur les mises et toutes les avances de quelque nature qu'elles soient. . . . Il n'existe rien de tout cela en Angleterre,

---

[1] Quoted by H. Sée, *La vie économique*, p. 46, n. 3.

[2] On the variations which affected the levy of tithes, see M. Marion, Article: Dime, *Dictionnaire des Institutions de la France, aux XVIIe et XVIIIe siècles*, Paris, 1923.

[3] In Weulersse, *op. cit.* vol. 1, p. 466.   [4] *Op. cit.* vol. III, pp. 264-8.

[5] Tocqueville, *L'Ancien Régime et la Révolution*, Paris, 1860, Livre II, ch. I, pp. 52-67.   [6] *Journal de Physique*, vol. XIV, 1778, p. 92, n. 2.

où l'opération est commune de faire son propre bien et de concourir à celui de l'Etat.'[1] Though this argument seems to come directly from the teaching of the 'Sect', in fact it is purely inspired by English example. François de la Rochefoucauld says precisely the same after his journey to England. 'La manière dont les taxes sont assises est ce qui donne le plus d'encouragement aux fermiers. Les taxes qu'ils paient sont lourdes, mais comme le prix de ce qu'ils vendent est en proportion, ils gagnent dans la même proportion. L'impôt auquel ils sont assujetis est sur les terres; s'ils trouvent le moyen d'augmenter le rapport de ces mêmes terres, ils ne paient pas plus.'[2] On the eve of the Revolution, Dupont de Nemours[3] clearly defined the new English-inspired doctrine: (1) Tithes should not affect any new crop. (2) The 'gros décimateurs'[4] should not claim extra tithes when the traditional cultivation is changed, as long as the previous product of tithes is sufficient for payment of 'portions congrues'.

England's example is similarly copied where State taxes are concerned. After a subtle criticism of the 'vingtième',[5] 'prime de découragement', Lazowski urges the settlement of fixed taxes. Then, the rise in prices of agricultural products would be pure profit for the peasants. 'L'impôt territorial en Angleterre . . . ce droit est absolument fixé pour les cultivateurs qui font valoir leurs propres terres. C'est le premier et principal encouragement qu'on donne dans ce royaume à l'agriculture, que d'affranchir du droit toute l'augmentation du produit qui est due à l'intelligence du cultivateur . . . que c'était à ce système que l'agriculture anglaise devait sa prospérité et ses progrès.'[6]

Another key to the agricultural question was the system of leases as it existed in the second half of the century. The agronomes

---

[1] *Idem.*

[2] François de la Rochefoucauld (1765-1848), *La vie en Angleterre au XVIIIe siècle*, Paris, 1945, p. 212.   [3] Pigeonneau et de Foville, *op. cit.* pp. 224 ff.

[4] Marion, Article: Dime, *Dictionnaire des Institutions. . . .*

[5] Pigeonneau et de Foville, *op. cit.* p. 331.   [6] *Ibid.* p. 396.

realized that it should be changed.[1] Dumont de Courset rightly pointed out that, 'si le fermier veut faire marner ses terres, comme il le doit, s'il veut semer du sainfoin, s'il veut enfin chercher à bonifier tous les terrains par la culture qui leur est propre, le terme de neuf années suffit à peine pour lui rendre les profits de ses frais.'[2] The necessity of longer leases, so that the new husbandry might be effected within a new agricultural framework, appeared vital to those who understood that the change was impossible, so long as the peasant was afraid of not being able to enjoy the fruit of his labours. François de la Rochefoucauld found that in Engand longer leases encouraged more efficient agriculture, and still more a class of well-to-do farmers, less ignorant and wretched than those in France.[3] He realized that the possibility of enriching themselves kept them attached to the soil they cultivated, by giving them a feeling of security and of ownership; whereas in France, if a man made a little money, 'fût-il même acquis par ce genre de commerce—ne voudrait plus rester fermier'.[4]

Thus the question of leases was of vital importance in agricultural improvement.[5]

An author, speaking of those who have their lands cultivated by tenant-farmers, advised them

de ne faire d'abord qu'un bail de six ans. Pendant cet espace de temps

---

[1] I shall speak here only about contracts between landlords and tenant-farmers, that is to say, those who cultivate the part of the feudal domain that the 'seigneur' exploits directly. I shall therefore leave out the question of the system of tenures and shall only speak of leases in so far as they were determined by the new character of agriculture.

[2] *Mémoire sur l'Agriculture du Boulonnais*, p. 177.

[3] 'Il me parait bien prouvé que, plus un fermier verse d'argent sur ses terres pour des améliorations, plus il gagne, mais pour cela il faut que son bail soit long parce que dans le cas où il serait court, il ne pourrait recueillir que la moitié du fruit de ses peines', *ibid.* p. 214.

[4] *Ibid.* p. 215. Also, De Pradt, *op. cit.* vol. I, p. 215, compares the mentality of English and French farmers. This was a reality: see M. Marion (article, quoted below p. 202, n. 3), 'Désertion des campagnes', pp. 352-3.

[5] Duhamel, *Eléments*, vol. II, pp. 384 ff.

la prairie se forme, le bétail se multiplie, ils construisent les bâtiments nécessaires et le fermier n'a point à se plaindre parce qu'il trouve lui-même tous les ans dans l'amélioration des terres et la multiplication des troupeaux, un nouveau profit sur lequel il n'a point compté. Ce premier bail étant fini, il peut en passer un de neuf ans. C'est alors qu'il recueille à plein, le fruit de ses peines et de ses avances.[1]

This is indeed a very optimistic and simplified view of the contemplated changes in husbandry. However, the question here is not to decide whether the proposed solution is right or wrong, but the outlook which closely linked the problem of farming on lease with that of the new husbandry. Once more it was apparently England which incited interest in long-term leases.

En Angleterre, l'usage a établi pour tous les biens fonciers, des baux à long terme, de 18 à 27 ans; par là le fermier, sûr de retenir les avances qu'il fait, ne néglige aucun des moyens d'engrais qui peuvent fertiliser le sol dont il se regarde en quelque sorte comme propriétaire. On pourrait considérer les Anglais comme un peuple de propriétaires cultivant eux-mêmes leurs terres; on sent l'avantage que l'agriculture de ce pays doit avoir et qu'elle a effectivement sur celle de France où l'usage a borné la durée des baux à neuf ans au plus.[2]

There was, finally, another problem on which no agreement was reached. This was actually the domain of the physiocrats, whose teaching on this point is famous; whether farming should be done on a large or on a small scale. The 'Sect' held for large scale farming, but their approach to the question was often more

---

[1] 'Prairies artificielles', *Nouvelliste oeconomique*, 1757, p. 78.

[2] *Mémoire de M. Le Sergeant d'Isbergue*, Archives du Pas de Calais, Etats d'Artois, Agriculture 9. Quoted by Calonne, *La vie agricole sous l'Ancien Régime dans le Nord de la France*, Paris, 1883, p. 64, Rozier quotes (Article: Bail, *Encyclopédie Méthodique*, vol. II, p. 19.) the 'Arrêt du 8 Avril 1762' which eases taxes on leases above 9 years and up to 27 years. Another advocated reform is that of covenants in leases (generally a strict maintenance of tradition) which was in open opposition with the new agricultural tendencies. See G. de la Fournière, 'Les Comités d'Agriculture de 1760 et de 1784', in *Bulletin du Comité des travaux historiques et scientifiques* (Section des sciences économiques et sociales), 1909, p. 115.

intellectual than practical.[1] In fact, the question was set by the agronomes in much more technical terms. Did the new husbandry require large or small holdings?[2]

Their answers varied. Very often 'the desire of the French experts was to create large farms of the English type',[3] but not always. The controversy which followed Duhamel's books is significant. Some said that such minute cultivation could be used only on small plots of land; others, that in order to diminish the cost, it should be tried on large surfaces.[4] But in this case convenient cultivation appeared of less importance than the cultivator's financial means, on which must depend the possibility of undertaking the new agricultural process. This gave rise to the long debated question of investment of capital in the land.[5] The existing state of agriculture allowed only those who possessed money to make the venture. Among these, two categories of individuals were particularly important—the rich landowner and the well-to-do tenant-farmer. In other words, the representatives of the new English agricultural classes, as they were organizing themselves.

L'Angleterre nous a donné le précepte et l'exemple ... l'Agriculture y a fait plus de progrès qu'en aucun pays du monde. ... Ajoutez à cela les mœurs des Anglais. Les grands et les riches résident une partie de l'année dans leurs terres; un très grand nombre de propriétaires

---

[1] This is one of the marked differences between physiocrats and agronomes. See, for instance, in Weulersse, criticisms of Mirabeau against Turbilly, a practical agronome, op. cit. vol. II, p. 163.

[2] H. Passy's book, Des systèmes de culture et de leur influence sur l'économie rurale, Paris, 1853, contains a discussion of the problem in the light of English theories of the eighteenth century, ch. I, pp. 9 ff. See also, H. Denis, Les prohibitions et les entraves à la libre exploitation des terres, Paris, 1911.

[3] L. Knowles, 'Economic causes of the French Revolution', The Economic Journal, 1919, p. 10.

[4] Wolters, op. cit. pp. 188 ff. Also G. Bourgin, 'L'Agriculture, la classe paysanne et la Révolution française', R.H.D.E.S. 1911, pp. 155-65.

[5] In Weulersse, op. cit. vol. I, pp. 379 ff. from the physiocratic standpoint. Especially see English-inspired arguments about lowering the rate of interest on loans.

opulents vivent toujours à la campagne; le Fermier et le Laboureur, en augmentant leur fortune, ne se dégoûtent pas d'un travail qui les enrichit sans les abaisser; une foule de Cultivateurs instruits et aisés observent par goût, par intérêt, par esprit de patriotisme; font des expériences sur les différentes propriétés des terrains divers, essayent de nouveaux engrais, simplifient et perfectionnent les instruments du labourage, excitent l'émulation parmi leurs voisins.[1]

The opinion that 'la grande culture', that is to say, 'tout entreprise de culture qui n'ayant qu'un seul exploiteur pourrait occuper trois ou quatre fermiers si elle était partagée entre eux' is the most productive one and can be more easily improved, can naturally be found among partisans of Tull's system. The improvement of agricultural implements, the purchase or manufacture of new ones, like drills, expenses entailed in enclosures; all this could be undertaken only by landowners or rich farmers who could afford it. It was so in England.

In 1780 Fréville published a *Recueil d'ouvrages sur l'Economie Politique et Rurale, traduit de l'Anglais*. The two volumes of the work included A. Young's 'Arithmétique Politique', Arbuthnot's 'Traité de l'utilité des grandes fermes et des riches fermiers' and his 'Essai sur l'Etat de l'Agriculture des Iles Britanniques'.[2] In the same year, the *Journal de Physique*, advertising the publication of the 'Essay on Agricultural Societies', criticized the French ones which did not encourage 'le travail en grand qui doit servir de leçon à tout un canton'. In short, not England's technique only, but its agrarian regime also, evolving into large-scale farming, was proposed.

Thus, in the agronomic writings, a sort of ideal agrarian state was taking shape. The great uncultivated estates were condemned, and their suppression proposed. Instead, the landlord would let to tenant farmers large areas of land, on which a more intensive

---

[1] 'Essays on Husbandry', *Gazette Littéraire de l'Europe*, London, 1784.
[2] A review of the book adds that the names of the authors can testify for the repute of the work, *Journal de Physique*, 1780, vol. XVI, p. 81.

and modern agriculture would be practised. He, or the tenant farmer, would provide the necessary money. But the great obstacle was that without it improvements were impossible.

This was, in a sense, the most extreme teaching derived from English principles, and is to be met with most frequently, at the very end of the period, when English results and English doctrines (mostly Young's) had led to interest in large farms.[1] It seems unnecessary to mention that *métayage* was condemned *a priori*, especially since its doom had been foretold by Young.[2]

The agronomes gave technical reasons for large-scale farming. This form of agrarian distribution was one of the solutions they offered for problems of fodder and cattle-raising, so very important in the new agriculture. The use of horses for cultivation (a symbol of the *grande culture*) was closely connected with the question of fodder, since the horse produces the straw it feeds on, whereas the ox requires ever more grass. Work was therefore to be done by horses, when cattle breeding was to become an important element of the modernized farm. Hoeing of the Tullian ridges, besides, required horses.

But even among the partisans of the new husbandry, unqualified imitation of England and of its large estates was not undisputed.[3] Many an author accused the anglomanes of not taking into account different geographical conditions, of structure and climate, and also certain well grounded customs. Mirabeau, for instance,

[1] *Essai sur l'Etat de la Culture Belgique et sur les moyens de la perfectionner* (Anonymous), London, 1784: 'révolution bien funeste au peuple, à l'état, la manie des grandes fermes, s'est accréditée tout à coup.' Section I, 'Nécessité de la réduction des grandes fermes', p. 25. On creation of large farms see a good account in G. Lizerand, *op. cit.* pp. 152 ff.

[2] *Travels*, vol. I, pp. 91, 98, *passim*.

[3] Certain writings did not favour at all the creation of large farms. The Chevalier d'Eon writes in a report to the French Government in 1763 about the agricultural transformations of England, 'on peut conclure de cet exposé, qu'il n'est point de pays où les enclos ne doivent se faire avec beaucoup de circonspection, et qu'il ne faut en aucune manière permettre le monopole destructeur des grandes fermes' in *Les loisirs du Chevalier d'Eon*, Amsterdam, 1775, vol. VII, p. 64.

opposed as he is to theoretical agriculture, praises small-size cultivation when it is appropriate to the land concerned. He gives interesting instances of husbandry near Marseilles in a country, the physical structure of which only allows small-size cultivation.[1] Others held absolutely opposite views about the best way of carrying out the method of Duhamel. Where some thought that only large-scale farming could make use of it, others wrote, 'les soins de détail que cette culture exige ne conviennent pas à de grandes exploitations'.[2]

However, on one point there was agreement. It was the role of the 'propriétaire'. The more moderate agronomes considered him as setting a good example rather than as an active reformer. Indeed, he could himself, by investing money in his lands, himself take the leadership of a whole region. As a rule, however, he was considered by Duhamel and his followers as a sort of patriarch. Thus, there was no need for the existing agrarian conditions to be changed at once; indeed, nearer the end of the century there was a violent counter movement against the large-scale cultivation theory. Small-sized land property was preferred by some, to large-sized property. Cliquot-Blervache, although he is an enthusiastic partisan of English husbandry, gives one of his chapters the title, 'De l'inconvénient des trop grandes propriétés'.[3] He writes in regard to the nobility's large estates, that they should be divided, 'en une infinité de petites propriétés libres, qui deviendraient d'autant plus libres entre les mains des nouveaux Franc-Tenanciers, qu'elles seraient plus partagées'.

In fact, many a landlord, however well acquainted with England, favoured small-scale farming in practice.[4] The most striking

---

[1] *L'Ami des Hommes*, vol. I, p. 55.

[2] Article: Culture, *Encyclopédie Méthodique*, vol. III, p. 717.

[3] *Essai sur les moyens d'améliorer en France la condition des Laboureurs, des Journaliers, etc. par un Savoyard*, 1789. See also Dumont de Courset, *op. cit.*

[4] Loutchisky, *La propriété paysanne en France à la veille de la Révolution*, Paris, 1912, pp. 199 ff. and Conclusion.

example was the Duc de Liancourt, a friend of the physiocrats, and an agronome himself, who received reports from his son in England, telling him of the prosperity of certain large farms in Norfolk,[1] and yet divided his own estate among small tenants, covering the hillside of Liancourt with cultivation, the variety of which made it look like 'un immense jardin'.

Should we discern a sign there of the anxiety of the nobility to increase their income by any means possible? It seems scarcely likely, for people like Liancourt had well-established fortunes and the increase could have been much greater in their income from their divided farms, had these *gentilhommes agronomes* attempted to follow closely the system advocated by their anglophile fellow-agronomes. We cannot deny that they were genuinely interested in agricultural problems, as Liancourt, for instance, made costly alterations in his estate. It may be, then, that they were following the other tendency, favouring small-sized farms, on which, with smaller investments, cultivation could be more prosperous. The author of *De la Culture Belgique*[2] expressed this point of view when he wrote:

Si on vient ensuite à vanter les grands avantages du riche fermier, à imaginer de nouveaux systèmes de culture, ou à redresser les pratiques de l'ancienne, à tenter de nouvelles expériences, on trouvera que tout cela ne conduit en effet les gens les plus aisés que sur le chemin de leur ruine, et qu'ils sont toujours à la fin forcés de revenir aux anciens usages, fondés sur l'expérience de tous les siècles. . . . On ne peut cependant disconvenir que des expériences de culture faites en petit . . . ne soient quelquefois bien avantageuses à l'agriculture, mais le petit fermier est à la portée de risquer quelque chose pour ces expériences comme le riche fermier.[3]

[1] F. de la Rochefoucauld, *op. cit.* pp. 219 ff.   [2] *Op. cit.* see above, p. 98, n. 1.
[3] *Ibid.* pp. 35, 47, 48. It is to be noticed that the Belgian agronomes chiefly were opposed to large-scale farming. They compared the Walloon districts where it existed and the produce of which was very low, with districts of small-scale cultivation between Ghent and Antwerp, the 'pays de Waes et Termonde', where agriculture was better understood and produced more. See article by M. de Burtin in *Actes de l'Académie impériale et royale de Bruxelles*, 1792.

This, from a man who has no great belief in the new husbandry. But it indicates clearly the position of the question at the end of the *ancien régime*. A total establishment in France of the English methods would have meant too violent a revolution in a country of which the social structure was so different from England's. It was therefore in an intermediate position that the agronomes stood at the end of the period.

Moreover, the form of peasant property as it had evolved through the centuries, was a major impediment to the establishment of large-scale farming. The existence of actual peasant ownership of the land, the infinite variety of its disposition in the rural districts and its transformations in passing from hand to hand by inheritance, had so entangled the long field strips of Northern France, or the small *compartiments* of the South,[1] that the creation of large divisions of the land in order to introduce the English husbandry would have meant the deepest changes in rural society;[2] so much so, that none even dared proffer general advice on the matter.

Duhamel, whose husbandry required large fields, understood the evils of a too great subdivision of the land. Once more he had recourse to the English example in speaking in the *Eléments* against a practice 'qui cause dans certaines provinces un grand obstacle aux progrès de l'agriculture' since, if the holding is too small, the cultivator is 'forcé de suivre la routine du pays et d'agir servilement comme ses voisins'.[3] A certain degree of animosity at 'la trop grande subdivision des terres'[4] made itself felt. All spoke of the agreements reached in England on the exchange and

[1] R. Dion, *Essai sur la formation du paysage rural français*, Tours, 1934, ch. II, pp. 34-72, clearly explains why the main regions to be reformed were the North, East and South-West of France. The agronomes generally belonged to these regions, which they would bear in mind when speaking of structural changes.

[2] Tessier pictures vividly the old agricultural landscape, 'Un pays partagé en une multitude de Propriétaires semble n'avoir qu'un seul maître', Article: Clôture, *Encyclopédie Méthodique*.   [3] *Eléments*, vol. II, pp. 163 ff.

[4] *Eléments*, Livre XII, p. 376. Also, *Jour. Oecon.* February, 1763.

re-grouping of lands,[1] but those who made the venture were few, and too many conditions were required for the changes to be carried out.[2] In addition, there was a good deal of protest to be contended with, so that the question was only posed and never solved.[3]

There is now a last question to be examined. How is it that a system which so obviously appealed to the rich, and allotted so important a role to the landowner, was not embraced by all the landowning classes? In other words, what are the relations between the agrarian changes proposed and the feudal reaction as it is called?

Even among theorists, the actual application of such methods to France seemed difficult in the extreme. Rich landowners, or members of the landed nobility, therefore, would have been rash (unless they were very rich) to undertake a revolution on the chances of which even the theorists themselves seemed not to agree. In spite of the genuine interest raised in those circles by the agricultural question, in spite of the increasingly marked tendency among the liberal nobility to model itself on the English aristocracy, no one in France dared use the latter's methods.[4] The French nobility, although it seemed at the time to be a heavier burden to the country than its English equivalent, did not in fact possess any real means of action, since the Government, although

[1] John Laurence, *New System of Agriculture* (1726), and Edward Laurence, *The Duty of a Steward to his Lord* (1727), were known in France at the time. The arguments proposed by the French are the same as those presented in Lord Ernle, *op. cit.* pp. 150, 151, 157, 161.

[2] In 1770 the Academy of Metz proposed the question for one of its prizes but it was unsuccessfully answered (Morogues, *op. cit.* Article: Réunions, vol. XVI, p. 324). See text of the question in Weulersse, *op. cit.* vol. I, pp. 365-6.

[3] In connection with this latter problem was the demand for a general survey of the land holdings in the kingdom. Here again we find allusions to England, 'C'est peut-être à l'égalité de cette fixation que ce Royaume doit l'augmentation de sa culture et de son peuple', Lazowski, *Essai sur l'Agriculture*, 1755, p. 371.

[4] On the success of the 'squirearchy', see Montague-Fordham, *A short history of English rural life*, London, 1918, p. 132.

it tended to favour them, did not for a moment forsake entirely the interests of the humblest member of the rural world. The nobility was both the judge and the defender in innumerable rural affairs, but it certainly played neither part when it was a question of the country's complete readjustment. Thus, in spite of its constant need for money, the means it employed to increase its income were limited. Instead of a real change on a large scale, only special and irritating decisions were taken, which were both a proof of weakness and showed a lack of understanding of wider problems. For the nobility of the end of the eighteenth century had never been trained to contemplate such problems as those which had long faced the British aristocracy; and only a small fraction of it could boast of any knowledge of political and social affairs. The 'realism' of the larger part of the nobility did not extend much beyond petty considerations of self-interest, or a stiffening of feudal taxation. In refusing to take part in the new economic evolution, the French nobility condemned itself to the political Revolution. An enlightened minority, the importance of which should not be overlooked, did indeed exist. But agricultural regeneration was a long-term undertaking. It did not pay immediately, and the followers of the agronomes, besides rousing the suspicion of the lower peasant classes, were incapable of setting a general example. Interesting as they were, their activities were only sporadic.

Throughout this debate, one element was always omitted in the agricultural books, even the most moderate. Wholly absorbed in the landowners, *personnes opulentes*, or rich farmers, they totally ignored the great mass of the yeomanry or small-holders and agricultural labourers. 'On n'a voulu apercevoir l'objet que sous trois faces et l'on a oublié la partie la plus importante, celle des manouvriers qui compose à elle seule les trois quarts de la nation. Le prix de leur journée n'a point haussé et l'avide fermier les a tenus dans une plus étroite puissance.'[1] Enclosures, the regrouping

[1] L. S. Mercier, *L'an 2440*, London, 1771, pp. 148-9.

of scattered holdings, extension of farms, all this meant the conversion of the peasant proprietor into a paid agricultural labourer, since such changes required an increase in manpower. In England this unhappy section of the agricultural world was to be sacrificed. The agronomes willingly accepted such a sacrifice in France, since it appeared necessary for the improvement of agriculture.[1] Blinded, as Marc Bloch says, not only by the benevolent and optimistic theories of the time,[2] but also by the prestige of English agriculture, they saw no harm in substituting 'une foule de propriétaires mal aisés à des salariés aisés'.[3] Duhamel and La Rochefoucauld-Liancourt held the same views as most members of the agronomic movement. Some, however, raised their voice against it. De Fresne, for instance, though a firm believer in the English methods, writing about partition of common lands, says, 'Dans ce partage des communes, il ne faudra pas oublier de réserver pour les pauvres plusieurs portions, dont on fera de temps en temps un nouveau partage. Les pauvres, comme les mineurs, ne doivent jamais perdre leurs propriétés; on doit toujours la leur rendre quand même ils l'auraient vendue.'[4] Thus theoretical and humanitarian considerations were mixed. None of them was to prove strong enough to succeed in effecting the revolution from outside. This was clearly shown by the failure of the Edicts on the partition

[1] The smallest change in the traditional methods meant a new handicap for the poorer peasants. In 1786 these complained against the use of the scythe as it deprived them of the stubble.

[2] 'Le fermier anglais retrace, plus que tout autre, le bonheur que l'imagination aime à attacher à cet état, et qui fut l'apanage du genre humain aux premiers âges du monde', De Pradt, op. cit. vol. I, p. 221.

[3] De Pradt also wrote, 'Les salaires régulièrement répartis par la grande propriété sont mille fois plus utile au bien être général, que la propriété personellement divisée en une multitude d'individus'. Ibid. vol. I, Avant-propos, p. xxxii. Léonce de Lavergne, Economie rurale de la France, Paris, 1866, Introduction, p. 22, holds the same views about the sale of church lands after 1789. The soif du sillon certainly prevented the capital from being invested in the rented land. See A. Young, Travels, vol. II, p. 605.

[4] De Fresne, op. cit. vol. III, p. 58.

of commons, which, in attempting to satisfy the big and small landowners dissatisfied both.[1]

It appears, then, that the English example is to be traced beneath the proposed solutions of the French agrarian problems.[2] The efficiency with which the English agricultural changes had been carried into effect by the firmness, and even harshness, of the landowners was still envied by the French agronomes at the beginning of the nineteenth century. Any proposed technical reform tending to an increase of agricultural production was bound to raise the mass of the peasant world against the big landowners. In fact, the former triumphed. The political Revolution did not, on the whole, much affect the agrarian division and rural character of France. After it, the anglomane agronomes, whose new god was Arthur Young, were still longing for 'l'enseignement des méthodes savantes et recherchées, qui exigent à la fois des études et des frais, des avances considérables en connaissances et en argent, telles à peu près que les méthodes employées par les agriculteurs anglais'. And De Pradt added, 'Quand la France possédera-t-elle une classe d'agriculteurs semblable à celle des Fermiers améliorateurs d'Angleterre, de ces hommes dont l'état est de rechercher les fonds négligés pour y porter leurs capitaux et leur industrie?'[3]

This sort of agricultural dictatorship was not to be. It is quite remarkable that although France gradually applied the main principles of the English technique, its natural consequences never followed. This lack of balance between theory and practice is to

---

[1] M. Bloch, *A.H.E.S.* 1930, *loc. cit.* pp. 527 ff.

[2] The question of grain legislation, though it was often discussed by agronomes connected with the Economic School, belongs to the history of economic doctrines and therefore is not to be treated here. The example of the English bounty for the export of grain was frequently advocated by economic and even agronomic writers. See Wolters, *op. cit.* p. 195.

[3] *Op. cit.* vol. 1, Avant-propos, pp. xliii-xliv. See his very interesting discussion about 'la grande culture' directly inspired by the teaching of Arthur Young, vol. 1, pp. 82 ff.

some extent still felt, and periodically threatens a practical solution, even nowadays. This shows that the English system remained an ideal without the possibility of complete application. The French countryside has kept its own peculiar characteristics, both good and bad. It never became, as the agronomes dreamt it might, another Norfolk or Suffolk.

PART FOUR

HOW THE NEW HUSBANDRY WAS INTENDED TO ENRICH FRENCH AGRICULTURE

CHAPTER VIII

THE NEW CROPS

The new husbandry was characterized by the variety of its crops as well as by their succession on the same ground. Among the new plants, some were said to have made the fortune of English agriculture. Others, although they were not new, needed rehabilitation and encouragement. The success of English farming, owing to the new crops, provoked a long and impassioned discussion. Agreement was reached on certain kinds of plants, interest in which had become so evident that they were no longer a subject of controversy. Against others, opposition was still violent at the end of the century, and sometimes the new plant could not overcome general indifference or prejudice. However, plants which were later to bring about profound changes in the order of cultivation—those which the peasant was to find it most profitable to cultivate—those in a word, which are advocated by the modern agriculture, made very noticeable progress.

In its first phase, the new husbandry was 'tout entière placée sous le signe des fourrages'.[1] Duhamel, like Tull, speaks of two kinds of fodder; one is 'raves ou rabes et gros navets'—roots; the other, herbaceous and perennial plants, like sainfoin and clover. These two types of fodder had their partisans and their opponents and it took nearly half a century before their advantages were generally conceded.

In 1750 roots were not absolutely unknown. In fact we find good descriptions of them in the old authors, but their use was very limited. Roots and turnips had their place in kitchen gardens

[1] M. Bloch, *Les Caractères Originaux*, p. 219.

109

and were mostly considered as vegetables and not as fodder. In 1750 their cultivation, although locally restricted, was no doubt common enough for Duhamel to write about it, 'suivant l'usage ordinaire'.[1]

He naturally contrasts it with the new husbandry, where the drill and the horse-hoe play an important part, in the chapter entitled 'Culture des navets suivant la nouvelle méthode'.[2] This was, in fact, the first important study about this crop to be published in France.

The *Encyclopédie* is rather vague on the subject. It points out, however, the importance of the new English fodder and its connection with sheep folding. The *Journal Oeconomique* continues to advocate the new ideas, in detailed and precise articles. Valmont de Bomare, a staunch advocate of English methods, also praises the new root and reiterates the established theory. 'Un arpent de terre semé de ces navets est d'un beaucoup plus grand rapport qu'en froment; d'ailleurs les racines divisent et préparent la terre à recevoir le blé et on recueille dans le même espace une beaucoup plus grande quantité de froment que dans une jachère ordinaire.'[3]

Roland takes up the discussion again, explains how the English use this fodder, and shows its importance in the achievement of English husbandry: 'En général on regarde cette découverte en Angleterre comme une des plus importante qu'on y ait faites depuis longtemps dans l'Economie Rurale.'[4] M. de Lormoy also says: 'Il est évident que l'Angleterre ne doit ses grands succès dans l'éducation des gros bestiaux et des bêtes à laine, qu'aux turneps.'[5]

For all that, the turnip was slow in getting recognition, more

---

[1] *Traité*, vol. I, ch. XIII, p. 155.
[2] *Traité*, vol. I, ch. XIV, p. 158.
[3] Article: Turnep, *Dictionnaire d'Histoire Naturelle*, vol. VI, p. 314.
[4] *Journal de Physique*, 1779, vol. XIV, p. 70.
[5] *Instruction sur la culture des Turneps ou gros navets*, Paris, 1786.

so even than other artificial pastures. All sorts of objections were raised, which are discussed in the same books. Turnips were alleged to be inacceptable to cattle and to require too much preparation for their consumption as fodder. Some critics advocated cutting them up; others leaving them whole, while others still insisted on the necessity of having farm-maids whose hand should be swift and arm thin enough to prevent cattle from being occasionally choked by a piece of turnip.[1]

Nevertheless, the acute shortage of fodders, sometimes resulting in actual disaster,[2] kept the question of turnips alive. The Royal Agricultural Society of Paris was much concerned with it. We find in its proceedings a memorandum on 'grosses Raves, Navets et Carottes', which indicates that 'l'usage de ces racines est si répandu en Angleterre ... qu'on ne saurait trop désirer de le voir universellement adopté dans les environs de Paris'.[3]

In December 1784 the Intendant of Paris published a pamphlet asking,

Est-on dans l'usage de cultiver en grand dans la Généralité, des Raves, des Navets ou des Carottes? En connait-on l'espèce appellée Turnip? Quelle est la quantité d'arpents qu'on emploie à cette culture? A-t-elle pour objet de nourrir les Bestiaux, ou bien vend-on les Turnips au marché comme les légumes? Sous quelle forme cette nourriture est-elle donnée aux Bestiaux et quel en est l'avantage?

The result of the enquiry shows that even the name 'turnip'

[1] *B.P.E.* 1786, vol. II, p. 106.
[2] 'L'extrême disette de fourrages qu'on a éprouvée en 1785 dans presque toutes les Provinces de France, a fait sentir aux Fermiers combien il leur était essentiel de perfectionner cette branche d'Agriculture qui a pour objet de rendre la nourriture des animaux plus abondante. Les effets de cette disette ont été moins funestes en Angleterre qu'en France, ce qui doit être attribué à l'importance qu'on donne depuis longtemps dans la Grande Bretagne à la culture des différentes plantes propres à la subsistance des animaux.' *Ibid.* 1788, vol. I, p. 36, Culture du chou navet par M. Arthur Young.
[3] *Mémoires d'Agriculture, d'Economie Rurale et Domestique publiés par la Société Royale d'Agriculture de Paris,* 1785.

was unknown in the *généralité*.[1] The report adds that, 'selon toute apparence on n'avait jamais vu de graine de Turnips d'Angleterre'. For this reason, the Intendant ordered the distribution of turnip seeds, recently purchased in England, together with an *Instruction sur la culture des navets, sur la manière de les conserver, etc. . .* which was to be completed in 1785 by the lengthy *Mémoire sur la Culture du Turnip* of Broussonnet, which summarized the contemporary knowledge of the subject.

The same claims were made by the other Agricultural Societies and especially by the Comité d'Agriculture which collected memoirs sent by others such as the Agricultural Society of Rouen. All praised the new crop very highly. Contemporary with the introduction of the English turnip were the efforts of the Abbé de Commerell on behalf of the mangel-wurzel, imported from Germany.[2] However, certain anglomanes tended to consider only the peculiarly English turnip. Lormoy writes that these 'sont d'un goût exquis, bien supérieur à celui des navets qu'il a vus en France. Il est donc intéressant de se procurer de la graine de ces turnips d'Angleterre, qu'on ne doit point confondre avec les raves, rabioules, etc.'[3]

On the eve of the Revolution, therefore, turnips seemed to be more an object of theoretical agronomic discussions than a cultivated crop of any significance. However, special importance must be attached to this controversy, even on the purely theoretical

[1] Abbé Rozier's indignation at this anglomania in terminology may be recalled here, 'Je ne sais d'où vient la manie d'introduire des noms nouveaux et de franciser des noms anglais, pour désigner des plantes qui sont connues de toute ancienneté en France. Le Turnep est le gros navet que l'on cultive de temps immémorial.' But this anglomania itself explained the favour given to the plant. The *Annales de l'Agriculture*, An. VI, vol. I, p. 332, say 'que le nom anglais de turneps ajoutait, ou plutôt était le grand mobile de l'opinion et du désir des cultivateurs de connaître cette plante.' A. Young definitely points out the distinction between the English turnip and the French *rabioule*, *Travels*, vol. II, p. 619.

[2] Pigeonneau et de Foville, *op. cit.* p. 22.

[3] *Mémoire sur l'Agriculture*, 1789, p. 103.

level.[1] England was also the home of another new fashion: the domestic cultivation of carrots. Its supporters quoted Miller who stated that 'un arpent de carottes peut nourrir plus de bestiaux que trois arpents de turnips'.[2] The results of Home's and Eliot's research on the same subject was also analysed and published.[3] The 'Société pour l'encouragement des Arts' advocated the intensive cultivation of carrots. Mr Billing's memoir on the subject was published with widespread success.[4] The extensive use of carrots as horse fodder in England was noted by François de la Rochefoucauld, in contrast to French custom of the time. 'L'on en donne partout aux cochons, même dans quelques parties de la France, mais nulle part aux chevaux.'[5]

The same happened with another root plant (as it was called in France in the eighteenth century) which has never lacked historians: the potato.[6] The fact has never perhaps been sufficiently stressed, that before Parmentier's famous campaign the potato was one of the principal elements of the *culture anglaise*. Its importance had been also noted in laudatory terms by several agronomers who were partisans of the new husbandry. Even if they were never as active as Parmentier, at least they drew attention to it, and kept it focussed on the new crop. In 1750, in the eyes of French agronomers, the tuber was specifically English. It was under its English name of 'potato' that it was introduced from England into France, 'On continue même à l'appeler ainsi dans toute la Grande Bretagne et dans quelques-unes de nos

[1] The importance of the beetroot perhaps came from the interest taken in the new kinds of vegetables; although its importance continued to be acknowledged, its cultivation spread very slowly until the middle of the nineteenth century; in Clapham, *op. cit.* p. 24.

[2] 'Manière de cultiver les Carottes en grand, pour la nourriture des bestiaux', *B.P.E.* 1787, vol. I, p. 35.   [3] *Ibid.* 1782, vol. I, p. 82.

[4] R. Billing, *An Account of the Culture of Carrots and their great Use in Feeding and Fattening Cattle, etc.*, London, 1765.

[5] Rochefoucauld, *op. cit.* p. 199. Though in 1789 M. de Guerchy 'ajoute qu'on commence en Normandie à imiter en cela les Anglais', *B.P.E.* 1789, vol. I, p. 127.

[6] F. Roze, *Histoire de la pomme de terre*, Paris, 1898.

provinces, en sorte qu'elle a été confondue et qu'on la confond journellement avec la patate et même le topinambour.'[1] After Duhamel's praise and the articles in the *Journal Oeconomique*, Valmont de Bomare devoted a very comprehensive article to it in his *Dictionnaire*. Having shown its importance in England and in Ireland, he adds, 'La culture de la pomme de terre . . . est digne d'attirer l'attention du Gouvernement et de chacun de nos cultivateurs modernes.'[2] In 1768 it was in Normandy, a region in which the new husbandry was making progress, that the cultivation of the potato was studied.[3] Ten years later Parmentier began his campaign. It seems, however, that the tuber was considered as a food for human consumption in particular regions only and in the enlightened classes of society.[4] The peasant world still considered it at the end of the century merely as an artificial pasture crop. In 1787 the Royal Society of Agriculture of Paris awarded a prize to M. Pierre Destouches, farmer at Villetaneuse, for his extensive cultivation of the potato, 'pour ses vaches'.[5]

The other sort of artificial pasture which was recommended was artificial grass. This is the most important kind of fodder. The problem of a sufficient supply of grass for cattle-feeding was the main concern of the agriculturists of the *ancien régime*, for it determined the possibility of corn cultivation. In spite of the opinions of certain writers who make the increased number of luxury carriages responsible for the shortage of fodder,[6] there is no

---

[1] Parmentier, *Traité sur la culture et les usages des pommes de terre, de la patate et du topinambour*, Paris, 1789, pp. 32-4.

[2] Article: Batatte ou Patatte, *Dictionnaire d'Histoire naturelle*, vol. I, p. 346.

[3] By the chevalier Mustel of the Rouen Agricultural Society, see *Mémoire sur la pomme de terre et sur le pain économique, lu à la S.R.A. de Rouen*, 1758.

[4] Wolters, *op. cit.* p. 229, quotes as propagandists for the new crop: Turgot in Limousin, duc de Broglie in Saintonge, M. Pauvilhiers in Poitou and Bertin's sister in Périgord.

[5] See *B.P.E.* 1786, vol. II, p. 129: 'Il y a déjà longtemps que les Anglais ont essayé de faire servir la pomme de terre à la nourriture des chevaux comme elle leur sert à celle des vaches, cochons, moutons.'

[6] In Pierre Jaubert, *Dictionnaire universel des Arts et Métiers*, 1773.

doubt the traditional husbandry was badly adapted to a possible increase of cattle. The new husbandry appeared to provide a source of fodders. In England certain artificial grasses had been rediscovered and used to better purpose. It was to be one of the tasks of the agronomes to publish their discoveries in this field.

After Tull, Duhamel had praised lucerne and sainfoin, the uses of which he found were not sufficiently well known. After him, Patullo developed the same theme at length and judiciously added some remarks of his own. He found that in France, hay was always mowed too late, so that it lost its colour, fragrance and nutritive value.[1] He advised cutting lucerne when it began to bloom before the stalks had had time to become hard and woody, and thus effectively replied to other objections to the new herbs.[2]

These two leading studies were the source of a very widespread movement of research on artificial herbs. Some were for sainfoin; others for clover or red clover. This last herb was the typical English grass, in the eyes of the agronomes.[3] In the North of the country peasants even kept its English name.[4] La Salle de l'Etang's books on artificial pastures was the first specialized treatise on the subject and enjoyed a lasting vogue throughout the century. Bomare's Dictionary states about the new grass, 'les

[1] Patullo, *op. cit.* p. 71.

[2] This objection is still met with in Gilbert and Cretté de Palluel. This defect 'faisait partie de ceux qu'avait la culture anglaise par planches, dite culture de M. Tull.' *B.P.E.* 1787. An excellent book (still praised in the nineteenth century) on the cultivation of lucerne and opposing the Tullian method was published by a Frenchman in England, Barthelemi Rocque, in 1761.

[3] 'Les cultivateurs flamands faisaient de temps immémorial à leur grand avantage un emploi fort étendu du trèfle, sans que nous l'eussions remarqué; mais les Anglais l'ayant apprecié et l'ayant apporté chez eux, nous l'ont fait connaître par leurs écrits il y a une cinquantaine d'années', Article: Trèfle, *Encyclopédie Méthodique*, vol. VI, p. 514.

[4] 'Le trèfle que nos paysans appellent clove, du mot anglais clover, trèfle', in Dumont de Courset, *op. cit.*

prairies artificielles sont regardées par tous les meilleurs agriculteurs comme un agent essentiel et même unique pour l'amélioration de notre agriculture.'

In 1777 the *Journal de Physique* drew attention to a new English book which was to provide further details on the matter;[1] and at the end of the period, Gilbert published his important treatise on artificial meadows. However, the views he expressed, although more precise and developed at greater length, are not much more original than those of Duhamel or Patullo thirty years before. The fact that he repeated them at many points well illustrates the difficulty the new crops met on their introduction into French ideas of agriculture. Besides, it must be noted that Gilbert had originally written his treatise for the Académie of Amiens, the centre of a country where agriculture was said to be more advanced than anywhere else in France.[2]

Besides this movement for the rehabilitation of certain plants, hitherto unjustly neglected, there was an effort to introduce into France some specifically English crops. One of these was ray-grass. The controversy over its virtues, its cultivation and its importance shows one of the most curious aspects of agronomic discussions in the eighteenth century.

First of all, apart from the general ignorance of its specific virtues, its botanical affinities were never absolutely determined in the eighteenth century. Duhamel, in his studies of artificial pastures, calls it 'fromental', 'dont les Anglais font un cas singulier',

[1] *The improved culture: Culture perfectionnée des trois principaux végétaux, la luzerne, la sainfoin, et la Pimpernelle, on y a joint des remarques concernant le trèfle*, London, 1777.

[2] Demangeon, *La plaine Picarde*, Paris, 1905, pp. 211 ff. P. Caron, 'Etat de l'agriculture ... dans la généralité d'Amiens, 1788', *Bulletin d'histoire économique de la Révolution*, 1909, pp. 121-39. Lefebvre, *Les Paysans du Nord, op. cit.* p. 206. The *agriculture flamande* does not seem to have deeply penetrated the country. In 1760 the *subdélégué* of Cambrai still admitted a lack of fertilizers in Cambrésis: 'Pour y remédier, on pourrait pratiquer les prairies artificielles comme en Angleterre.' On spreading of Flemish technique see Lefebvre, pp. 207-8. But it must be pointed out that formal examples are always alluding to England.

and complains that it is not sufficiently known.[1] Patullo calls it by its English name 'ray-grass' and praised its excellence when sown with clover. The *Journal des Savants*, reviewing Patullo's book, adds: 'Ce Rai Grass est ce que nous appelons l'yvraie, dont le mélange avec le bled est, comme l'on sait, si pernicieux à l'homme qu'il pourrait être une nourriture très nuisible aux bestiaux.'[2] Here begins the confusion between the new plant and tares (*fausse ivraie*, as it is often called) which was to be so prejudicial to its success. However, its advocates were not discouraged. The Society of Bretagne 'eut bien désiré de pouvoir cultiver le Rai Grass. Cette plante qui n'est connue que sous son nom Anglais, est citée par tous les cultivateurs d'Angleterre avec les plus grands éloges'.[3] It became the subject of a significant Memorandum which ends with a parallel between French and English husbandry.[4] At this stage the new crop became a sort of symbol of the new agriculture.

But this vogue was threatened by a pamphlet violently opposing the introduction of ray-grass.[5] The author, although he claimed that he had obtained his seeds from England, thought that dry ray-grass was prejudicial to horses, while it was too bitter when dry, and that, on the whole, cattle preferred clover. 'Il ajoute qu'il est fâché que tout ce qu'il a eu à dire du Ray Grass soit opposé à ce qu'en rapporte M. Duhamel.'

Fréron added some time later, not without melancholy, 'Je regrette tous les jours l'espèce d'oubli où est tombé le Ray Grass,

[1] *Traité*, vol. v, Préface, p. xviii.

[2] November, 1758, p. 40.

[3] L. de Villiers, *Histoire de la Société d'Agriculture de Bretagne*, p. 23.

[4] Prairies artificielles, *Mémoires sur le Fromental et la Culture anglaise* par Dom Miroudot, together with *Mémoire abrégé sur le sainfoin ou l'Esparcette tiré des recueils de la Société oeconomique de Berne, auxquels on a joint quelques remarques de Tull et Duhamel. Instructions sur la Culture de la Luzerne distribuée dans la généralité de Bordeaux*. Lyon, 1762.

[5] *Mémoire sur le Ray Grass et le Red Clover, par M. Delisle*, Paris, 1761 (memorandum read at the meeting of the Royal Agricultural Society of Paris, 2 July 1761).

cet excellent pâturage que le témoignage d'une nation entière et des expériences repétées dans notre climat n'ont pu soutenir contre le rapport d'un seul homme, trompé peut-être par des observations imparfaites.'[1]

The fifth volume of the *Gentilhomme Cultivateur* contains a complete defence of ray-grass, which was repeated by the Comte de Maupeou in a letter to Demportes.[2] But it was Bomare who explained the main reasons for the opposition to ray-grass. Although 'cette plante qui est cultivée en Angleterre et en Irlande pour former des prairies artificielles a toutes sortes d'avantages qui devraient nous engager à la cultiver', there had been a mis-understanding about the grass itself. Agricultural connoisseurs had only received rye-grass from England, 'c'est ce qui a occasionné des plaintes contre le vrai ray grass qui, dégénérant en mépris, ont entrainé le discrédit de cet excellent fourrage'.[3]

However, the critics never ceased to raise objections to the grass. Even in 1782 another memorandum attacked the authors who favoured it: Forbonnais, Duhamel, Dupuy-Demportes, Bomare, and Mortimer. After reckoning all the various expenses it involved, and the poor results it produced, the memorandum concluded: 'ce serait être dupe que de sacrifier son temps, ses peines, son argent et ses terres, à la culture d'un pareil végétal.'[4] There is no doubt that too much enthusiasm had been shown about the new artificial grass. But *a priori* criticisms were equally wrong.[5] Planazu pointed out that within certain limits of cultiva-

---

[1] *Année Littéraire*, 1762 (review of Desplaces' *Préservatif contre l'Agromanie*, vol. II, p. 63. [2] *Gentilhomme Cultivateur*, vol. v, p. 112.

[3] Article: Ray Grass, Fromental ou Faux Froment, *Dictionnaire d'Histoire Naturelle*, vol. v, p. 298.

[4] 'Mémoire sur diverses espèces de Plantes propres à servir de fourrage aux Bestiaux', par M. L. Clouet, *Journal de Physique*, 1782, vol. XXI, p. 338.

[5] Dumont de Courset wrote about it these suggestive lines: 'Dans tous les essais, sur vingt personnes qui en font, à peine y en a-t-il cinq qui réussissent parce que la vivacité naturelle des Français, qui n'ont en vue qu'une prompte jouissance, les éloigne des soins et de la constance qui sont nécessaires pour s'en procurer une durable', *op. cit.* p. 129.

tion, ray-grass might be advisable; and Cretté de Palluel, after Gilbert, introduced a sound distinction. The climate of England is better than that of France for this crop. In France it must be mixed in natural pastures with the other native plants. The agronomes never ceased to envy the admirable quality of English grass. However, their efforts were not entirely unrewarded. In 1785, an obscure *subdélégué* in Northern France informed his Intendant of 'quelques épreuves d'un petit graminée Anglais, qui, je crois s'appelle "régrave" '.[1]

If the acceptance of these was slow in practical farming, a better knowledge of other plants used in England was quickly spreading among theoretical agronomists. Here botanical research and husbandry were mixed, and the progress of the former helped in the development of the latter. In the elaborate memorandum of Louis Clouet we find a detailed description of several plants which can be used as artificial pasture, to which new importance was attached. Such were burnet, hop-clover, vetches, beans, peas, esparcet, timothy-grass, bird-grass and so on. These were henceforward classified, their particular interest was pointed out and, as a whole, they were to be included on a larger scale in French agriculture.

The Agricultural Society of Rennes ordered an enquiry about plants in natural pastures that cattle liked or disliked. Thus an incentive was given to an extensive movement for classification,[2] and improvements.

But as the spreading of English methods was slow, shortage of fodder was still sometimes acute. Here, again, inspiration was found in England, and the Baron de Servières printed a pamphlet on the preservation of leaves as a winter fodder, 'avec l'addition

---

[1] In Calonne, *op. cit.* 29 January 1786.

[2] Academy of Dijon 1779: 'Déterminer les plantes vénéneuses et utiles qui infectent souvent les prairies de Bourgogne et diminuent leur fertilité; et indiquer le moyen le plus avantageux d'en substituer de salubres et d'utiles, de manière que le bétail y trouve une nourriture saine et abondante.' We find the same concern in most of the books at the end of the century (e.g. Planazu, 'Des herbes qui doivent composer une prairie', *Oeuvres d'Agriculture*, 1778).

des moyens connus en Angleterre pour la conservation des feuilles en fourrage pendant l'hiver'.[1] The Royal Society of Agriculture ordered the publication of Arthur Young's *Memoir on Cabbage*,[2] a discovery which was to be enlarged by Sonnini de Manoncourt. The latter also published some good studies on the *radis* of Lapland[3] and Canadian lentil.[4]

Until the very end of the period there was an uninterrupted vogue for English fodder. In 1789 M. Cels, Receveur des Fermes du Roi, and afterwards a distinguished agriculturist, offered to order in London a new grass discovered by M. Thomas Walker, 'dans un temps où l'on parait (suivant le système d'Agriculture Anglaise) reconnaître la vérité d'augmenter les pâtures'.[5]

So far as corn was concerned, if the new husbandry evolved new ideas about its cultivation, they did not result in any deep change in the species. It may be noticed, however, that certain agronomes showed more interest in spring corn, the introduction of which was recommended in certain systems of rotation. Besides, new kinds of seeds had been discovered. The *Gentilhomme Cultivateur* pointed out the interest of a different sort of barley cultivated in the neighbourhood of Patney in Wiltshire, which he called 'l'orge de Patney' and which might be of interest because of its early crop. This is an example of the tendency to plant crops well adapted to the soil, and the agronomes looked for examples in England when such possibilities existed.[6]

[1] *Mémoires de la S.R.A. de Paris*, 1786. Also in *B.P.E.* 1786: 'Les moutons de l'Angleterre qui donnent la plus belle laine sont nourris avec la feuille d'orme; dans les provinces méridionales du royaume, on leur réserve pour l'hiver, les extrémités des peupliers et on en fait de petits fagots.' See also Pigeonneau et de Foville, *op. cit.* p. 51.   [2] *B.P.E.* 1788, vol. I, pp. 36 ff.

[3] *Mémoire sur la culture et les avantages du chou-navet de Laponie*, Paris, 1788.

[4] Quoted by the *Foreign Essays . . . 1765* as having been recently introduced by M. Dumesnil Coste in Normandy and being 'preferred to all other pulse for feeding horses and black cattle'. In 1789 the *Journal de Physique* advertised 'on trouvera cette graine chez Villemorin-Andrieux, Marchand Grainier fleuriste, Quai de la Mégisserie et Hotel de Calais, Rue Coquillière'.   [5] *B.P.E.* 1789.

[6] Sprat-barley was indeed popular in Wiltshire. In Moffit, *op. cit.* p. 4.

But the main value of the new husbandry in the question of corn was that it allowed the land to produce more and better corn. Duhamel had already said that the advantage of the *nouvelle culture* was that it prevented the land from being occupied 'par des grains moins précieux'. Therefore a selection of lands had to be made. He opposed the general practice in France which consisted in trying to grow corn by any means whatever. 'Les cultivateurs ne doivent pas s'entêter, et si leur terre n'est pas bonne pour le blé, qu'ils essaient . . . colza, oeillette, garance, gaude, etc. . . .' The grains of corn themselves must be as big as possible, and in order to attain this result, the seeds must be carefully selected. Gleaners' grain is particularly advisable as it has been chosen ear for ear. The *Essai sur le Commerce d'Angleterre*,[1] and the *Tableau de la Grande Bretagne*[2] testify to the quality of English corn, a consequence of its improved cultivation. This interest shown in the grain question had far-reaching consequences. Among those the most notable were the works of Tillet, Tessier and Duhamel on grain preservation and the fight against corn diseases; and those of Parmentier on corn flour. All these are important aspects of French economic history in the eighteenth century.

Wheat cultivation, then, became the symbol of a richer agriculture. A review on Patullo's book stated:

Il ne fait mention ni de l'avoine, ni du seigle, parce qu'il prétend que la grande quantité qu'on en sème en France est une marque certaine du mauvais état de son agriculture; et il avance avec raison que si on trouvait le moyen de faire porter du froment à toutes les terres en France, le peuple qui ne consomme actuellement que du seigle en serait bien mieux nourri, et plus en état de supporter la fatigue de son travail.[3]

Connected with the question of grain is that of straw which is so precious to farmers, as it can be used for fodder and litter as well as a fertilizer. Like other products of the new husbandry, the

---

[1] Enlarged from J. Cary's work (1695) by Butel-Dumont, 1755.
[2] Baert, *op. cit.* vol. III, p. 246.
[3] *Journal des Savants*, November 1758, p. 38.

quality of straw should improve. Mixed with lucerne and sainfoin it could make an excellent winter fodder. It was used in this way in England. 'Les chevaux pâturent en été dans le trèfle; en hiver ils ont de la paille d'orge et dans les gros travaux, on leur donne un bushel d'avoine par semaine et quelquefois on la mêle avec de la balle de grains . . . les cochons sont nourris . . . de chaume en automne.'[1] Besides, as field enclosures would prevent vagrant flocks from grazing on stubble, the peasants could, therefore, take it off and use it on the farm or plough it into the ground, where it made an excellent fertilizer before winter. Finally, the modernized process of storing corn was to destroy certain bad customs, such as leaving bundles of corn on the ground, so that the grain became puffed up with air moisture, which deteriorates both the grain itself and the straw.[2]

Tree cultivation also concerned the agriculturists in the eighteenth century, and several of its methods were again English-inspired. Before 1750 an important section of the Administration was already devoted in France to the maintenance of forests, and a tradition of interest in tree-plantations had existed since Colbert's time.[3]

As in the case of fodder, so too wood presented a shortage problem in the eighteenth century, as a result of slackness in the administration of *Eaux et Forêts* and indiscriminate tree-felling.[4] One of the chief reasons against breaking up new lands was that it tended to suppress woods and forests indiscriminately. Before 1750 Duhamel and Buffon had dealt with the question, and were

---

[1] Baert, *op. cit.* vol. III, p. 247.

[2] This custom ('javeller' oats) especially was universally condemned at the time.

[3] The reform of the administrative system of the *Eaux et Forêts* was made in 1669 (*Ordonnance générale*). Few changes occurred in the eighteenth century. See *Loix Forestières de France . . . par M. Pecquet*, Paris, 1753 (review in *Journal des Savants*, May 1754, pp. 184 ff.).

[4] H. Sée, *Histoire économique*, p. 200. M. Rouff, *Les mines de charbon en France, 1744-91*, Paris, 1922, pp. 11 ff. See 'Ecole des arbres forestiers', *Journal de Physique*, 1789, vol. XXXV, pp. 317-19.

chiefly known as specialists on timber. Like Buffon, Duhamel first studied vegetable physiology, and criticized and completed the works of Grew, Stephen Hales and Ellis.[1] Here again, Duhamel gave the impetus to the study of arboriculture in France. His followers were numerous. Their first aim was to restock forests with wood of good quality which might be used for different purposes: copse-wood for heating, straight timber for large-scale building.[2] The regeneration of the French navy in the second part of the eighteenth century led to a new interest in timber, and English treatises on this question were studied.

Besides all this, there was a marked effort made to naturalize in the country certain foreign or exotic species, a movement which imitated what was being done at the same time in England.[3] Here it is possible to trace the influence of Miller, the gardener of Kew and Chelsea, whose works were often translated and always highly esteemed.[4] He is called in a letter published in the *Journal de Physique* (1778) 'le premier jardinier de l'Europe parce qu'il tenait un rang chez un peuple qui cultivait les arbres les plus rares lorsque nous pensions à peine à multiplier les plus communs'. It is possible then to find there, to a large extent, the influence of gardening *à l'anglaise* which required a greater variety of trees, differing in shades, aspect and so on.[5]

In the third place, the question of trees was also determined by the question of enclosures. Amoreux, who wrote a good book

[1] T. Thompson, *op. cit.* p. 45. See Duhamel, *Des semis et plantations d'arbres*, Paris, 1760.

[2] Bomare, Article: Chêne, *Dictionnaire d'Histoire Naturelle*, vol. II, p. 66. Goyon de la Plombanie was a very active researcher on timber.

[3] Like the cultivation of the cedar tree after advice by M. Lawrence, 'savant anglais'; Bomare, *op. cit.* vol. I, p. 631.

[4] *Traité des arbres résineux conifères, extrait et traduit de l'anglais de Miller . . . par M. le Baron de Tschudi*, Metz, 1768.

[5] L'Héritier, *Sertum Anglicum, seu Plantas rariores quae in hortis juxta Londinium, imprimis in horto Regio Kewensi excoluntur*, Paris, 1788. Le Blanc had already noted the abundance of exotic trees in England and the efforts made in nursery gardens to multiply them.

about hedges,[1] devotes part of his work to the study of trees used for this purpose, and often quotes English discoveries.

In short, in this question of trees it would be going too far to say that English influence was conclusive. However, it can always be felt. It is either an indirect or a complementary influence. Its result was the introduction of new kinds of trees and a better understanding of the use of different woods. Many a tree which is now familiar to our eyes, like the Italian poplar,[2] owes its existence in France to the great movement of research which was inaugurated after the agricultural revival.

Finally, in the less important spheres of agriculture, it is also possible to acknowledge to some extent the influence of English methods. In gardening and the cultivation of industrial plants, they frequently offered a new conception, or indicated a better technique.

In the eighteenth century the achievements of English gardening had thrown into the shade the old French supremacy in this field.[3] The number of English books on gardening shows that 'il doit être mieux entendu ici qu'ailleurs'.[4] There is no question here of pleasure gardens.[5] It was market gardening which was so greatly admired in England. Treatises on gardening by Miller, Bradley, and the 'Jardinier de Milord Robert Manners', were translated and widely circulated. The smallest details of the cultivation of plants and their recipes filled the pages of the equivalent French books. Bradley's book[6] is an 'ouvrage rempli d'observa-

---

[1] *Mémoire sur les haies . . . par M. Amoreux fils*, Paris, 1787.

[2] Pelé de St Maurice, *L'art de cultiver les peupliers d'Italie*, Paris, 2e ed. 1767.

[3] The French school of gardeners or 'jardinistes' as Père d'Ardène said (1769) must not, however, be underestimated. It included François le Gentil, Dahuron, Le Berryais, Schabel, La Bretonnerie, Butré, etc. . . . all famous in their time and estimable authors.   [4] Le Blanc, *op. cit.* Letter XLI.

[5] On 'Jardins à l'anglaise' see Lerouge, *Jardins anglais et chinois*, 1776-89; Blaikie, *Diary of a Scotch gardener*, London, 1931; F. C. Green, *Minuet*, London, 1939, pp. 246, 247, 248, etc.

[6] *Observations physiques et pratiques sur le jardinage et l'art de planter, avec le calendrier des jardiniers, Ouvrage traduit de l'Anglais de Bradley*, Paris, 1764.

tions curieuses et de préceptes excellents, dictés par l'expérience devient utile et même nécessaire à tous ceux qui font leur occupation ou leur amusement de la culture des terres et des jardins'.[1] At the end of the century the books of English gardeners known in France were extremely numerous, such as those by John Hill,[2] Thomas Burnes, Mawe and Abercrombie, the latter of which 'est destiné à faire connaître les erreurs, à corriger et à suppléer aux omission des Ecrivains qui les ont précédés'.[3]

What then are the results of English gardening which interested the French? First of all, the variety of its products, especially the fine quality of its green vegetables. Secondly the art with which luxury or exotic products were grown. The cultivation of pineapples in greenhouses was an English speciality, like the cultivation of asparagus and all kinds of fruits and vegetables, imported from the colonies.[4] The cultivation of flowers for display, which was encouraged by the tastes of amateurs of gardening, was also an English achievement.[5] Careful attention to detail as well as great expenditure on gardening implements, were the reasons for this superiority. This provided an obvious contrast to the backwardness of French gardening at the end of the eighteenth century. The impetus given by La Quintinie had not been sustained. Complaints of this can be found in contemporary books: 'on se plaint dans les Provinces de l'extrême difficulté d'y trouver des

[1] *Année Littéraire*, 1756, vol. VI, p. 26.

[2] *Eden or a Compleat Body of Gardening*, London, 1757.

[3] *Journal de Physique*, 1779, vol. XXXV, p. 318. 'The Universal Gardener and Botanist. Thomas Mawe, Jardinier du duc de Leeds, et M. Jean Abercrombie.'

[4] Bradley was the first to advertise in France a practical way of growing these luxury fruits and vegetables. Rozier quoted Miller's *Treatise on the Ananas*, 1769. See curious remarks in Grosley, *Londres*, Lausanne, 1774, vol. I, p. 127; and Adam Taylor, *A Treatise on the Ananas*, 1769. Thouin, in the proceedings of the Paris S.R.A., wrote about rhubarb, 'cette plante est déjà cultivée avec succès dans différentes parties de la Grande Bretagne', 21 July 1785.

[5] It was after a visit to Kew Gardens that Mme de Genlis brought back to France the first samples of moss-roses: A. Britsch, *La Jeunesse de Philippe-Egalité*, Payot, Paris, 1926, p. 414. In 1775, they were a rarity in France. See *Lettres de l'Abbé Morellet à Lord Shelburne*, Paris, 1898, p. 67.

jardiniers instruits de bons tailleurs ou conducteurs d'arbres. Les riches propriétaires sont forcés d'en faire venir à grand fraîs, de la capitale.'[1] Here again a very conservative spirit existed. Besides, English gardening was considered, and with reason, to be purely a luxury item, fashionable only in the richest circles. For the rest of the country local practices, like Montreuil gardening, remained standard examples. However, this English influence must not be altogether neglected. Quite apart from new techniques in planting and grafting, French gardens came to acquire new plants, hitherto uncultivated.

Nearly the same situation existed in regard to industrial plants. This cultivation was in a way a particularly French tradition which was not easy to alter, as its products were used by a state-controlled industry whose regulations were extremely conservative. In spite of all this, discoveries and inventions from across the Channel were received with interest and some of them began to have a place in the French agronomic system. The translation of the Essays of the Dublin Society, besides important articles on breweries, contains a most significant essay on flax cultivation, 'qui intéresse particulièrement la Bretagne'.[2] The cultivation and preparation of dye plants, such as madder, were also affected by the teaching of English methods.[3] But this cultivation was too limited and too traditional for noticeable changes to be introduced.

Such were the main aspects of English influence on the kind and quality of cultivated crops in France.

[1] *Journal de Physique*, 1789, 'Ecole pratique pour l'Education des Arbres Forestiers, pour les Pépinières et la taille des arbres'.

[2] The Society of Brittany complained about the preference given to flax imported from the Baltic countries.

[3] *Méthode de Cultiver la Garance, telle que les Hollandais la pratiquent en Zélande . . . à laquelle on a ajouté la méthode de cultiver la Garance en Angleterre. Par M. Miller, Membre de la Société Royale de Londres*, London, 1758. Also 'Moyen de conserver la racine de garance sans la dessécher, publié a Londres' (*B.P.E.* 1785, pp. 332 ff).

CHAPTER IX

THE PROBLEM OF LIVESTOCK

THE problem of cattle was very urgent in eighteenth-century France. A chronological study of agricultural literature would show that cattle problems were more and more studied from 1750 onwards. Between 1750 and 1770 problems of soil cultivation preponderated. After 1770 the number of books on cattle increased and connections with English methods grew progressively closer. Although reasons were rather different from those which gave rise to the same movement in England, the success of English experiments had a very great influence on the French problem.[1] It must be in fact pointed out that, while in England researches were directed towards an increment of meat 'gained by better methods of breeding directed specifically towards that purpose',[2] propaganda was made in France in favour of an increment of the number of cattle. Indeed, the question of improvement was also raised, but it was the number of animals that could be supported with the new agriculture which seemed to matter mostly.

The importance of cattle had for long been recognized; Liger wrote, 'on peut dire que le bétail à cornes fait la richesse de la campagne et que c'est de lui que nous vient cette abondance de

[1] The connection between the new husbandry and success in stock breeding appeared immediately to French observers. De Lormoy wrote, 'L'Angleterre en est un exemple frappant; ce royaume doit ses succès à cette culture, tant pour l'amélioration des terres que pour la multiplication des gros bestiaux et des bêtes à laine; elle met les cultivateurs à même d'avoir toujours au moins une demi année de fourrages devant eux', *Mémoire sur l'agriculture*, 1789, p. 76.

[2] G. E. Fussell, 'Size of English cattle in the eighteenth century', *Agricultural History*, vol. III, no. 4, 1929, p. 100.

toutes choses que nous y voyons régner'.[1] Chomel also pointed out the urgency of the cattle question. France had not enough cattle and the existing breeds were more or less degenerate.[2] But as these early agriculturists emphasized mostly the need for draught purposes and manure, and as the consumption of meat was relatively small,[3] they showed no marked interest in the improvement of breeds in so far as the question of meat or dairy products was concerned. However, Chomel, in well-informed articles about English sheep and goats,[4] declares that the process which gave the English their beautiful breeds should be imitated, for it was the problem of wool, an urgent requirement of French industry, which he had in mind.[5]

After 1750 the question, while remaining fundamentally the same, was enlarged in scope by the introduction of the new theories on husbandry. The connection which existed between the 'new system' and the problem of livestock was substantially to amplify the latter, so that it was of capital importance on the eve of the Revolution in 1754. An agronome summed up the whole problem by saying:

Les progrès qu'il [the French] a fait dans les Arts et les Manufactures prouvent assez que si son goût le déterminait pour la culture des terres, il retirerait de ses campagnes infiniment plus de secours qu'il n'en reçoit à présent et que, bien loin d'être obligé de recourir à ses voisins pour avoir bien des choses dont il manque actuellement, il se trouverait bientôt en état de les leur fournir lui-même à un prix raisonnable. Tels sont les chevaux de toute espèce, les laines, le beurre et en général toutes denrées que donnent les bêtes à laine et à cornes dont les

[1] *Op. cit.* ch. XVII, p. 145.

[2] Still the conclusion of De Pradt in 1801, *op. cit.* vol. I, pp. 151, 209 ff.

[3] On the consumption of meat by the peasants see Babeau, *La vie rurale*, pp. 104, 105.   [4] Article: Chèvre, *Dictionnaire oeconomique*, vol. I, p. 600.

[5] Already in the seventeenth century, travellers noticed that English sheep 'portent de la laine fine dont nous voyons à Paris des étoffes et des draps que nous appelons draps d'Angleterre, aussi beaux que s'ils étaient de soie. . . .' Jouvin de Rochefort, *Le Voyageur d'Europe.* . . . Paris, 1672, 2nd part of vol. III, quoted by Ascoli, *op. cit.*

Français seraient munis abondamment, s'ils voulaient s'occuper du soin d'élever ces animaux et de leur procurer les nourritures convenables.[1]

The agronomes struggled against this state of things and turned their eyes to England in order to find out how the English obtained their results in breeding cattle, horses and sheep.[2]

The necessary food was to be obtained through the intensive cultivation of artificial pastures. The respective merits of lucerne, sainfoin and turnips and their importance for fattening animals or increasing their production of milk are always described. But chapters on selection or reproduction, or on the improvement of breeds are rare. In short, it seems that on the question of cattle, the French agronomes actually believed that their 'livres économiques . . . paraissent ne laisser rien à désirer sur cet article'.[3]

However, two questions held their attention: English methods of breeding and the fight against cattle disease.

The latter one became more important with the growth of agricultural knowledge. And just as a consequence of the new agriculture was a greater concern about corn diseases, so the new interest shown in cattle questions was to lead to the elaboration of veterinary medicine.[4] Fight against cattle plagues had been in progress since the end of the seventeenth century,[5] and, as a

[1] 'Projet sur la manière d'élever les chevaux', *Nouvelliste oeconomique*, 1754, vol. III, pp. 115-16.

[2] Pigeonneau et de Foville, *op. cit.* pp. 327 ff. Since the seventeenth century English cattle had achieved its repute of excellence; see Ascoli, *op. cit.* p. 290. At the end of the eighteenth century, F. de la Rochefoucauld still wrote, 'Le bétail est le fondement de leurs fermes, et la manière dont ils l'entretiennent, le principe de leur richesse', *op. cit.* p. 220.

[3] 'Sur le choix et le gouvernement des vaches', *Journal oeconomique*, April, 1752, p. 31.

[4] See Paul M. Bondois, 'La protection du troupeau français au XVIIIe siècle. L'Epizootie de 1763', *R.H.E.S.* 1932, pp. 352-75. Very little is found in Sir F. C. Smith's *The Early History of Veterinary Literature*, London, 1915-23.

[5] See list of Treatises on cattle plagues prior to 1750 in the Catalogue of Huzard's Library, Paris, 1842, pp. 229 ff. Especially, *Réflexions sur les Maladies des Bestiaux qui règnent à présent avec les remèdes pour les traiter: imprimées par ordre du Maréchal Duc de Villeroy*, Lyon, 1714.

consequence of the new movement of interest in increasing live-stock, a whole school of veterinary surgeons appeared. The creation of the Veterinary Schools of Lyons and Alfort, one of the finest achievements of the *ancien régime*, can in a way be regarded as an indirect consequence of English influence on French agriculture. Mostly concerned at first with hippiatric studies the field of their activities was soon extended to cattle and they obtained several successes in their fight against cattle plagues. Some of the curative methods used were closely connected with English methods. Inoculation[1] was one of these and we find in one of the most widely read books of the period the method of a gentleman of Yorkshire who attempted successfully, 'une sorte d'inoculation pour préserver les bestiaux de la suite de la con-tagion'.[2] Indeed, the French agronomes were aware that cattle diseases were not unknown in England. But they were more frequent in France because of 'la trop grande domesticité, la malpropreté, la mauvaise nourriture, l'ignorance et les préjugés'.[3] Also, as the century progressed, the changes brought about in English livestock had their repercussion in France. Lord Ernle speaks of this time as of a time 'when beef and mutton were to be more necessary than power of draught or fineness of wool'. Beef was, in fact, considered as the traditional food of the English people.

'Point de bœuf, point d'Anglais', wrote St Amand. Although the urgency of condition and taste was less imperious in eighteenth-century France, greater care was being advocated in the produc-tion of meat and even milk and cheese.[4] M. Guerrier, brother of

---

[1] The *Journal Oeconomique* says it was discovered in Pembrokeshire and 'used everywhere in Wales'.

[2] Valmont de Bomare, Article: Taureau, *Dictionnaire d'Histoire Naturelle*, vol. VI, p. 105.

[3] Roland's memorandum, in *Journal de Physique*, 1779, vol. XIV, p. 71.

[4] 'An Essay on the best method of making butter, on the Management of a Dairy . . . by Mr Jore, Secretary to the Society of Agriculture, established at

M. de Lormoy and a great admirer of England, obtained by improved methods products of superior quality. He wrote to his brother, 'Notre crème en ce moment est aussi délicieuse que celle du mois de mai et aussi parfaite que celle de Londres.'[1]

Therefore better methods of breeding were necessary. In France, the lack of enclosures prevented cattle being kept in the open without a keeper; so they were fed in sheds. That is why French farms had so many buildings. The English, on the contrary, had only the minimum of buildings. Cattle were left unsheltered without suffering any ill effects. These peculiarities of 'l'éducation sauvage des bestiaux' were studied by several agronomes, and Roland gives a very careful account of the methods used in England.[2] The opponents of the principle of the 'éducation sauvage' were relentless and we shall give an account of their opposition when discussing the question of sheep breeding. But the minute technical descriptions of the anglophile agronomes helped to a better understanding of breeding methods.[3]

English influence was much stronger in the question of horse-breeding. This was a traditionally French activity, but its origin and methods were partly outside the problem of agriculture.[4] Royal studs had been established in the seventeenth century, but the strict observance of the regulations which should have governed

Rouen in Normandy'. (*Foreign Essays*, 1768.) See also articles by Tessier, 'Lait', 'Fromage', in Rozier, *Encyclopédie*, etc. Musset, in *Le Bas Maine*, Paris, 1917, quoting Sanson's *Traité de Zootechnie*, vol. I, p. 2, says that until 1850 there is no evidence that farm animals were even considered but as auxiliaries for vegetable production. H. Sée, *Histoire économique*, p. 190, speaks of cattle in the eighteenth century as a 'necessary evil'. These views are certainly much exaggerated.

[1] *B.P.E.* 1787, vol. I, p. 103.

[2] *Journal de Physique*, 1779, vol. XIV, pp. 63 ff.

[3] In *B.P.E.* 1785, vol. I, p. 119 ff. which summarizes the methods recommended by the Society of Bath on calf-rearing without milk. Also in 1786, those of Young on the same subject.

[4] Mostly connected with military problems. See important Bibliography in Huzard's *Catalogue*.

them was neglected.[1] In 1750 French horses, and especially those used for agricultural work, were generally animals of mediocre quality whose breed needed improvement.

In the middle of the seventeenth century there arose a new interest in horses.[2] Very soon, under the influence of English horse-races, a kind of hippomania established itself in aristocratic circles and the interest in horse-races was keen at the end of the period.[3] Equally important was the direct imitation of England in the use of horses in rural life. All the travellers remark that horses are common amongst the agricultural community.[4] Even the smallest farmer owned one. The French were amazed to see country people riding on horseback. At the turn of the century English horses still provoked the same admiring descriptions.[5]

This movement in favour of cultivation with horses, these eulogies of English horses, however understandable, arose from predisposition towards English agriculture, considered as the type of large-scale agriculture. That is why it was so enthusiastically adopted by the physiocrats.[6] In agronomic literature, these views were sometimes opposed, often toned down. And as small

[1] See Musset, *Histoire de l'élevage du cheval en France*, Paris, 1917 ('L'Administration des haras en France au XVIIIe siècle'). *Revue d'Histoire Moderne et Contemporaine*, 1909-10, vol. XIII, pp. 36-7, 133-52. Also *Eloge de J. B. Huҙard*, by M. Pariset in vol. I of Huzard's *Catalogue*. Interesting remarks on the traditional understanding of French horse-breeding can be found in De Pradt, *op. cit.* vol. II, pp. 150 ff. ('Des Chevaux').

[2] Partly as a consequence of the wars in the middle of the century 'Les chevaux Anglais passent pour les plus hardis et les plus propres à la guerre', Butel-Dumont, *Essai sur le Commerce d'Angleterre*, Paris, 1755.

[3] Patullo, *op. cit.* p. 271, 'En effet, c'est l'émulation des prix distribués par toute l'Angleterre aux courses des chevaux, qui porte ses haras au point que les races qui en sortent sont recherchées par toute l'Europe.'

[4] François de la Rochefoucauld, Le Blanc, Mme du Boccage, Grosley.

[5] Baert, *op. cit.* vol. III, p. 251; Grosley, *Londres*, vol. I, pp. 29, 309, 321 ff.

[6] In E. Daire's edition of Quesnay's Article: Des Fermiers, *Economistes français*, Paris, 1846, vol. I, pp. 221-8. Baudeau, *Introduction à la philosophie économique*, vol. II, pp. 694 ff.

holdings had their energetic champions, so also had cultivation with oxen.[1]

Nevertheless, the necessity of improving the existing breeds was never denied. Numerous books were written on this subject. For a long time the standard books on horses, besides what was found in the *Maison Rustique*, were those written by the Duke of Newcastle and Solleysel.[2] After 1750 the question was reviewed in its entirety. Writers urged the imitation of English methods and tried to find places in France suitable for extensive breeding. Treatises on hippiatrics were continually being published until the end of the century by men like Lafosse, Dupuy-Demportes,[3] Chabert, Bourgelat, Huzard and the pupils of the Veterinary Schools. On the question of state-controlled studs opinion was divided.[4] Should the government still control them, or was it better to give the peasant more freedom to cross his horses as was done in England? At the end of the *ancien régime* some people attempted to arouse interest in horse questions by means of horse-races and exhibitions of purebreds in *comices agricoles*.[5]

English breeding was advocated as a means of success. The horse then came to be considered as an animal of value, whose beauty and quality could not be paid for too highly. Therefore a

[1] de Sutières, 'Comparaison du cheval et du bœuf pour les travaux de la campagne, spécialement pour les labours', *B.P.E.* 1789: 'Les Anglais, plus attentifs à leurs intérêts que nous . . . ont déjà dans plusieurs provinces substitué les bœufs aux chevaux.' However, cultivation with horses tended to spread. H. Sée, *Histoire économique*, vol. I, p. 196.

[2] *Méthode et invention nouvelle de dresser les chevaux par le . . . prince Guillaume, Marquis et Comte de Newcastle . . . seigneur de Cavendish*, traduit de l'anglais, Anvers, 1668, other editions in 1672, 1674, 1677, 1700, 1737. J. de Solleysel, *Le parfait maréchal*, 1st ed. 1664, regularly reprinted and translated until 1775.

[3] Adapter of J. Bartlet in his *Le Gentilhomme maréchal de l'Anglais de J. Barthelet*, Paris, 1756-8. Translations of English works on the question were not lacking. One of the most celebrated was Garsault's *Anatomie générale du cheval traduite de l'anglais de Snap*, Paris, 1733-7.

[4] E. Cavailhon, *Les haras de France*, Paris, 1886-9.

[5] For instance, M. Le Boucher de Crose, Member of the Société de Bretagne, author of a *Mémoire sur les haras*, Paris, 1770, organized horse-races.

careful selection was necessary. 'Tous les ans, dans les Annonces et les Gazettes Anglaises, on propose à louer des étalons vaillants et beaux pour couvrir les juments. On demande jusqu'à 10 guinées et même plus pour chaque femelle qui sera couverte; prix exorbitant sans doute. Ferait-on cette dépense si on ne savait par expérience que l'excellente race qui en provient remboursera au delà les avances que l'on a faite de se la procurer?'[1]

Horse food had to be carefully selected. The finest straw often chopped, hay, oats, beans, and carrots were the principal fodder given to horses in England. Stables and litter should be clean and well aired. These English methods were in obvious contrast to the neglect and routine methods denounced by Tessier in a district like the Sologne, for instance.[2] Nevertheless, the impulse given was to have definite results. Some members of the nobility possessed studs equipped in the English manner.[3] They bought certain of the finest specimens of English thoroughbred horses. Although this may seem a restricted field of application, the influence of such a fashion was strong. English bloodstock found its way into the studs of Normandy and the fact that some of the qualities of French horses were finally to disappear roused concern about this English invasion. The beginning of the nineteenth century was to see a reaction against this fashion.[4] But the English example had,

---

[1] *Journal de Physique*, vol. I, Introduction.

[2] After his official inspection in Sologne after 1766. See E. Menault, *Souvenirs de Beauce, Cassegrain, Blanchet, Tessier*, Paris, 1859, pp. 67-111.

[3] Robert Black, M.A., *Horse Racing in France*, London, 1886.

[4] On the craze about English horses see Amédée Britsch, *La jeunesse de Philippe-Egalité, 1747-85*, and *Lettres de L.P.J. d'Orléans à N. P. Forth, 1778-85*, Paris, 1926.
The reaction was expressed in very violent terms. Linguet already wrote, 'Les courses de chevaux semblent se soutenir en France; les appréciateurs sensés doutent cependant qu'elles y produisent jamais les bons effets qu'on leur attribue en Angleterre. Ils prétendent qu'elles n'exciteront au delà du Pas de Calais qu'un enthousiasme frivole et ruineux. Elles engagent les Anglais à nourrir, à élever des chevaux, à perfectionner les races qu'ils naturalisent chez eux' (*Annales politiques, civiles et littéraires...*, London, 1777, p. 181). De Pradt, in spite of his

at any rate, helped to make the French agronomes realize the importance of horse breeding and a real improvement in its methods was very noticeable at the very end of the *ancien régime*. But on the whole, the most important question among those raised by the problem of livestock was that of sheep. Literature on this subject is considerable, and among the writers were men like Daubenton or Tessier who acquired a fame which places them in France on the same level of learning and genius as Bakewell in England. Their influences on government circles led to definite experiments and action;[1] finally, the improvement of sheep was in the process of becoming a reality at the outbreak of the Revolution. The fact that this intense movement was almost exclusively determined by the example of England has never been sufficiently emphasized.[2]

For the French agronomes the important point in sheep-breeding was to obtain better wool. The question of meat was only secondary for them. Their main motive arose out of the incentive given by Colbert and his successors in the seventeenth century to the French wool industries and their intention to counterbalance

enthusiasm for English methods, bitterly complains, 'Les chevaux anglais ont envahi les haras de Normandie. . . . Au lieu de ces anciens chevaux normands remplis de vigueur et d'un entretien si facile, on n'aperçoit presque plus que ces longs et maigres chevaux de modèle anglais, espèces de squelettes. . . .' (*op. cit.* vol. I, Avant propos, p. lvii). The same opinion was held by Huzard *le fils* ('Notice sur les chevaux anglais et sur les courses en Angleterre', in *Mémoire d'Agriculture*, vol. xx, and 'Rapport fait à l'Académie Royale des Sciences le 22 Septembre 1817, sur la précédente notice'). Huzard the father considered that 'dès 1789 l'anglomanie avait eu parmi nous des résultats plus généraux et plus funestes. Elle détériorait nos haras, elle brisait notre industrie, elle détruisait nos chevaux. . . . Il n'est pas jusqu'à ces courses dont elle nous avait donné le goût et qui n'étaient et ne seront peut-être jamais que les variétés d'un luxe onéreux et trompeur'. Notice . . . by M. Parisot, *op. cit.*

[1] Bibliography in Levasseur, 'Des progrès de l'agriculture française dans la seconde moitié du XVIIIe siècle', *Revue d'Economie Politique*, 1898.

[2] Even though it was the Spanish merino sheep which finally was most used for cross-breeding, English technique of breeding originated the movement. The fact is not pointed out in Sir John Clapham, *The Economic Development of France and Germany 1815-1914*, Cambridge, 1936, p. 25.

the success of English manufacturers in this domain, mostly in weaving fine or half-fine woollen cloths. Chomel had already expressed this feeling of competition with English methods. Everybody knows, he said, that one of the reasons for England's wealth is the flourishing condition of its wool industry. He tells how, in the fifteenth century, English breeds were crossed with Spanish ones and described the regulations about sheep preservation existing in England.[1] This proves the existence of an interest in cross-breeding which had been, so far, alien to French methods.

The *Encyclopédie*, the physiocrats and Duhamel himself were agriculturally rather than industrially minded. So they have no views on this point. But the question was raised again with greater vehemence as the new agriculture assigned a special role to sheep. Besides, the movement was encouraged by the economists, who were great admirers of England and well connected with ministerial circles.

In 1754 we find Colbertian arguments in an article on sheep: if France raises better sheep, it will not need to spend money in foreign countries on the purchase of wool. 'L'Angleterre en est un exemple bien capable de nous encourager.'[2] The same idea was expressed frequently. The Academy of Amiens gave as a subject for a dissertation in 1754 the question: 'Si on ne pourrait pas se passer en France des laines étrangeres.'[3]

In the year 1754 other works on sheep were published. One

---

[1] Article: Brebis, *Dictionnaire Oeconomique*, vol. I, pp. 436 ff. On 5 February 1728 the Comptroller General le Pelletier sent a memorandum on sheep breeding to the Intendant of Dauphiné, in which he said, 'Il serait à désirer que l'on put exciter les principaux fermiers et gros laboureurs qui nourrissent des troupeaux . . . à pratiquer les mêmes soins et la même attention [as in England], et il n'est pas douteux que si quelqu'un d'eux y réussit, il n'inspire la même émulation à plusieurs autres.' In Babeau, *La Province sous l'Ancien Régime*, Paris, 1894, vol. II, p. 240.

[2] 'Profit sur la manière d'élever les chevaux. . . .' *Nouvelliste Oeconomique*, 1754, vol. III, p. 120.

[3] *Année Littéraire*, 1755. Also, G. Bourgin, 'L'Agriculture et la Révolution', *R.H.D.E.S.* 1911, p. 164. Weulersse, 'Le mouvement préphysiocratique', *R.H.E.S.* 1931, p. 258.

of them, after criticizing the crossing of sheep from Languedoc with sheep from Berri, suggests that foreign breeds should be imported.[1] This point of view was emphasized a year later by the publication of Swedish treatises translated into French, which indicated the best means of improvement, and it is again the English example which, after a detour through Sweden, was proposed.[2] The Chevalier Alström had crossed his sheep 'après avoir considéré que les Anglais ont gagné plusieurs centaines de millions en annoblissant leurs brebis par l'introduction des béliers d'Espagne dans leur pays'.[3] In 1768 Bomare's Dictionary agreed that there was an urgent need of improvement in French breeds. Once more the example set by England was proposed. 'De semblables exemples ne doivent-ils pas nous animer?. . . Que l'on multiplie dans le Cotentin . . . l'espèce des bêtes à laine d'Angleterre; la nature des pâturages, la disposition du lieu, tout annonce qu'on y recueillera une laine pareille à celle des plus belles toisons d'Angleterre.'[4]

Besides the question of wool there was also the question of manures. Sheep, as well as cattle, were an indispensable element of the new husbandry because of the excellent quality of their manure. The *Encyclopédie* had already pointed out the importance of sheep in English agriculture and the English method of sheep folding. Volume VI of Duhamel's *Traité* devotes one complete chapter to sheep and gives advice about folding them. At the same time, another agronome spoke highly in praise of sheep raising in

[1] 'Mémoire sur les moyens de bonnifier les laines dans les Provinces du Royaume', par M. le Marquis de Puismarais, *Nouvelliste Oeconomique*, 1754, vol. III, p. 109.

[2] 'Instruction sur la manière d'élever et de perfectionner les bêtes à laine', par Fr.W. Hastfer, traduit du Suédois par M.xxx. Paris, 1756. See *Journal des Savants*, January 1757, pp. 65 ff.

[3] *Nouvelliste oeconomique*, 1755, vol. I, p. 55. The book of Jonas Alströmer is a treatise, in Swedish, on sheep management (1720). It seems to have been rather popular in France, as it is possible to find it quoted until the end of the century. Alströmer was well acquainted with English methods from which he borrowed, after a journey to England, 1719-20 (see notice in *Svenskt Biografiskt Lexikon*, vol. I).   [4] Article: Bélier, *Dictionnaire d'Histoire Naturelle*, vol. I, p. 376.

enclosures.[1] His precepts were obviously inspired by the theory of Patullo. In demonstrating the significance of enclosures in the usual way, he assumed the objections that they spoiled the wool and did not suit the animals.[2]

In 1760 the theory of the improvement of French wool as well as the theories of artificial pastures, enclosures, and 'éducation sauvage' was definitely worked out.

In 1778 M. de Mante published a book, of which the title itself indicated the connection between the 'new husbandry' and English sheep breeding.[3] But at the same time, the *Journal de Physique* published a lengthy article by Roland de la Platière, a most significant work in which the whole of the new theory was contained.[4]

Roland tried to explain why the quality of French wools was much inferior to the English. He drew a very detailed and accurate picture of the methods commonly used in the Provinces. Here he criticized the *bergeries*, where the animals lived in filthy conditions, there he noticed that no use was made of the sheep-fold; the small and degenerate French breeds not only produced mediocre wool but the management of the wool itself led to considerable loss in weight and quality.[5] In comparing French and English fleeces he obtained the following result.

[1] 'Manière avantageuse de nourrir les bêtes à laine', *Nouvelliste Oeconomique*, 1754, vol. II, p. 130.

[2] M. de Brou, Intendant of the Rouen *généralité*, made experiments on sheep folding from 1751 to 1760, which 'left no room to doubt, but that the sheep folded abroad were more healthy and stronger than the others, and that their fleeces were finer, softer and whiter', *Foreign Essays*.

[3] *Traité des prairies artificielles, des enclos, et de l'Education des moutons de race anglaise*, Paris, 1778.

[4] 'Mémoire sur l'Education des troupeaux et de la culture des Laines par M.R.D.L., Inspecteur général des Manufactures de Picardie', 1779. See also his article on wool manufactures in the *Encyclopédie Méthodique*, Section 'Commerce'.

[5] In Artois and Boulonnais fleeces were washed after shearing but also after a kind of fermentation which made cleaning easier but gave the wool a yellowish colour. In other Provinces where wool was sold by weight, before cleaning, the flock was taken along dusty roads so as to increase the weight of the fleeces by dust and impurities.

	Fleece before cleaning	Fleece after cleaning	Last and perfect cleaning
Boulonnais	6 lb.	3 lb.	2lb. 4 oz.
England	6 lb.		5 lb. 1-10 oz.

This bad management of sheep in France was a factor of diseases which too frequently attacked the animals. Thus the proportion of their number to the surface of soil on which they lived was comparatively small, whereas in England 'on compte les moutons par millions'.[1]

England, indeed, appeared as a sheep's paradise:[2] 'En imitant les Anglais dans leur pratique, on obtiendra les mêmes resultats qu'eux.' The food must be proportioned to the volume of the animals and certain breeds could not be raised anywhere. Besides, the open life in pastures, 'l'éducation sauvage', was one of the most important features of English sheep management.[3] Even in counties without enclosures, sheep lived freely in the open. The picture Roland draws of his arrival in England and his first acquaintance with English sheep is a very picturesque one and reflects his surprise on encountering this new method of breeding.

Lorsque j'arrivai en Angleterre, je fus jeté sur la plage à trois heures du matin à quatre ou cinq milles de toute habitation et j'errai dans les

[1] See the estimates given by E. G. Fussell and Constance Goodman in 'Eighteenth-century estimates of British sheep and wool production', *Agricultural History*, vol. IV, no. 4, 1930, pp. 132-3. Figures vary from 11,000,000 to 43,000,000. [2] Roland visited Kent and Sussex.

[3] Huzard, in the Introduction to Daubenton's *Instruction pour les bergers*, points out that in 1750 experiments had already been made on 'l'éducation des bêtes à laine à l'air libre, dans la vue d'améliorer les laines'. This had taken place in the park of Chambord, under official supervision and had entitled their owner, M. de Perce, to special privileges and a favourable report. See *Arrêt du Conseil d'Etat du Roi, concernant le régime et l'éducation sauvage des bêtes à laine*, 15 August 1752. Also in Rozier's Article: Bergerie, *Cours d'Agriculture*.

prairies pendant plus de deux heures sans rencontrer figure humaine. Mais elles étaient couvertes de troupeaux. Ce fut pour moi un spectacle assez intéressant que la grosseur, l'embonpoint de ces animaux, leur blancheur éclatante, leur air étonné et fugace semblable à un troupeau de biches.

He praises, on the other hand, the selection of rams and choice of covering time which produced stronger lambs, born in milder weather and when the grass is growing. Particular care in the preservation and improvement of the sheep's body attracted his attention. After the castration of lambs, their tails were cut one month later. 'Les Anglais prétendent que cette opération carre l'animal, lui arrondit la croupe, fortifie ses différentes parties et lui donne plus de dispositions à engraisser.' Cleaning fleeces with soap kept them in a remarkable state of whiteness and smoothness. This perfection in the production of wool explained why it was 100% cheaper than in France, while manpower was 25% and lands 100% more expensive than in Boulonnais. He also points out one of the natural advantages of England: the absence of wolves. In France, it was due to the fear of wolves that peasants kept their sheep in sheds. At the time when Roland was writing, the memory of the Gévaudan beast was still fresh in men's minds. A memoir published before Roland's work insisted on the importance of enclosures in keeping wolves away, and gave as an example Normandy, where there were enclosures and no wolves.[1] The whole of England enjoyed this condition, which was one of its advantages over France.[2]

Roland's reflections are very interesting, not only because he

---

[1] Article: Cloture, *Encyclopédie Méthodique*, vol. III, p. 317. See *Méthodes et projets pour parvenir à la destruction des Loups dans le Royaume, par M. De Lisle de Froncel*, Paris, 1768.

[2] In the seventeenth century, the fact that England was without wolves had already been pointed out. Some attributed it to a 'natural antipathy', others to methodical destruction, or quality of English dogs. La Fontaine wrote in 'Le loup et les bergers': 'C'est par là que de loups, l'Angleterre est déserte. On y met notre tête à prix.'

gives a vivid account of the methods used in England,[1] but also because throughout his work he compares them with French methods and holds them up as an example to his compatriots. His position as Inspector of Manufactures undoubtedly enabled him to have a precise knowledge of the situation in France and therefore his attitude towards English models must be considered as completely devoid of any *a priori* prejudice in their favour.

But it seems that Roland was too severe on traditional French customs. He was to be attacked on this count by an agronome, who reached almost identical practical solutions, but had different tendencies: the Abbé Carlier.[2] In two lengthy articles in the *Journal de Physique* (1784)[3] the latter takes up again the problem of French wools and refers constantly to England. But his conceptions are appreciably different from those of Roland, whom he often contradicts. He is a typical representative of the tendency to react against foreign, and especially English, influence. He challenges the opinion that English flocks are the best in the world,[4] and says that such a claim is to be found only in the works of French writers. 'Les auteurs Anglais les plus enthousiastes n'ont jamais formé une telle prétention.' He also opposes the fashion of importing English rams, advocated by Roland and Daubenton. Why this admiration for English breeds? 'Ce qui vient de loin passe ordinairement pour avoir un mérite plus distingué que ce qui croit sous nos yeux. La rareté fait le prix de bien des choses; l'éloignement et la privation rendent plus vifs les

---

[1] His picture should be compared with that of E. G. Fussell's in 'Animal Husbandry in eighteenth-century England', *Agricultural History*, 1937, vol. XI, no. 3, pp. 189-207.

[2] A specialist on the question of sheep and wools; wrote several treatises and books on the matter in 1754, 1756, 1762, 1763. Travelled in the kingdom in the official capacity of Government Inspector from 1764 to 1768.

[3] 'Mémoire sur les moyens de perfectionner les laines de la France', and 'Observations historiques sur l'état ancien et l'état actuel des troupeaux et des laines en Angleterre', *Journal de Physique*, 1784, vol. XXIV, pp. 271-80.

[4] He criticizes the *Etat du Commerce d'Angleterre*, 1755, vol. I, p. 33.

désirs d'avoir en sa possession des objets dont on a une idée extraordinaire.' Were these lines prompted by his bitter jealousy of Roland[1] or were they the reaction of common sense and sanity, coupled with a genuine belief in the virtues of French farming? Carlier's critical attitude is not without value. It is true to say that at the time when he wrote, English sheep were far from having attained the qualities which were to make the Bakewell and Coke breeds famous. But he went too far when he refused to admit that English breeding methods were superior to those of France. He agreed that there were some undesirable features in sheep management: badly drained meadows, the right of common pasture, and closed stables, 'où l'air ne pénètre pas'. But he refuses to accept the principle of open-air sheep rearing, for fear of wolves and robbers. Carlier considered also that sheep manure was almost lost in enclosures. He preferred the shed system which allowed manure to heap up, in readiness for its immediate use by the peasant. His arguments against the English method are interesting, although they are in the general line taken by its opponents. According to him, the French farmer was not interested in growing wool. Wool was generally bought very cheaply and there were too many middlemen between producer and factory. Besides, even when sold directly to the manufacturer, wool was badly paid. More than wool, French peasants needed manure. 'La production des grains obtient nécessairement la préférence sur celle des pâturages. C'est le cas de tirer d'abord parti des bêtes à laine pour l'amendement des terres et de pourvoir ensuite à la culture des terres.' This is in the best conservative agricultural tradition. But his comparison between the English and French understanding of agricultural problems, based on a different economy of the country, is not lacking in perspicacity. It was easy for England to sacrifice its

[1] The tone of his polemic is very incisive: 'Il est donc évident que la méthode anglaise modifiée par M. Roland de la Platière doit être regardée comme l'un de ces systèmes qui s'échappent des cabinets avec l'éclat d'un bruyant artifice et en ont le sort.'

corn cultivation to more extensive pastures, and to concentrate on the problem of wool. 'Les Anglais, à la faveur de l'immense étendue de leur commerce maritime, tirent de toutes parts une quantité de grains et de denrées de première nécessité qui les dispensent d'ensemencer une partie de leurs terres. Ils trouvent dans cette ressource la facilité de convertir en prairies naturelles et artificielles, des portions considérables de terrains dont on abandonne les herbages à de nombreux troupeaux.' So he agrees that some of the English principles must be followed, especially those concerning improvements in stabling and the care taken of animals, but on the whole the English method of sheep breeding could not be applied in France.

Carlier is not the only representative of this tendency. As usual, the opponent of close imitation of England had his followers. Some of them tried to oppose anglomania with arguments used in England itself. Such was M. de Lamerville who advocated keeping sheep in sheds and found his examples in Gloucestershire.[1] As we reach the end of our period, writings on the question of English breeding become more numerous. Their authors were all experimental farmers, equally well acquainted with the new theories. They all claimed to have obtained the best results with or without imitation of England. Until 1790 new books were published on the subject almost annually, by the Marquis de Lormoy, Quatremère d'Isjonval, Broussonnet, Carlier and Roland.[2]

But the definitive work on the question was to be written by Buffon's collaborator, Daubenton, who, in publishing his *Instruction pour les bergers*, produced one of the most celebrated treatises of the eighteenth century.[3] This work is a sort of corpus of all the

---

[1] M. de Lormoy also opposed the method of the sheep-fold and found his reasons in England itself, *Mémoire sur l'Agriculture, op. cit.* p. 47.

[2] A list of memoranda on the question of sheep can be found in Levasseur, 'Des progrès de l'Agriculture'. Besides, the controversy fills the greater part of the *B.P.E.* from 1782 onwards.

[3] *Instruction pour les bergers et les propriétaires des troupeaux*, Paris, 1782.

knowledge recently acquired about sheep and it most vividly describes the old French technique of sheep management. Sheep anatomy and physiology were carefully studied and also their modes of living and their food. Questions of cross-breeding, management of flocks, and the education of shepherds with specialist knowledge were treated. The book may be considered as one of the best examples of agricultural literature in the eighteenth century and the climax of an evolution in agricultural writing.

Verified by personal experiments, English precepts were firmly established and Daubenton successfully combined certain French provincial customs and the latest innovations from England. His influence was to be considerable.[1] At the end of the Revolution, Daubenton was actually considered as the leading authority on sheep. He had, besides, enlarged the results of his *Instructions* with the addition of valuable works published under the protection of the government which, under the influence of men like M. de Tolozan, was taking the greatest interest in the new experiments in sheep crossing and wool production. The aim of Daubenton thus became to obtain wools of equal quality to those of England. In the research station he set up, he tried to improve his breeds, and attained the success he had predicted, 'si nous prenons de l'émulation comme les Anglais pour faire valoir nos troupeaux.' A series of memoirs on the quality of wool he had grown and the clothes made from it, proved that France could compete successfully with England in this field.[2]

In 1785 the question of sheep was much in vogue. That year witnessed a dearth of fodder. Immediately the government entrusted Daubenton with the promulgation of instructions to

---

[1] Huzard wrote in his Introduction to the *Instruction* (An. x), 'elle éveilla aussi l'attention des propriétaires et l'amélioration fit des progrès assez rapides.'

[2] *Mémoire sur le premier Drap de Laine superfine du cru de la France, par M. Daubenton*, Paris, Imprimerie Royale, 1784. *Addition au Mémoire précédent*, 1784. *Observations sur la Comparaison de la Nouvelle laine superfine de France*, 1785.

prevent the death of a great number of animals.[1] The government's instructions insisted on the advantages of sheep-folding, and contained new technical details borrowed from England and the pays de Caux. In 1788 an enthusiastic follower of Daubenton, a farmer and an active member of the Royal Society of Agriculture, the Marquis de Guerchy, began an intensive drive to combine Daubenton's precepts with his own personal English experiences.[2]

Finally, at the very end of the period, Bakewell's experiments began to be known. He, in fact, represents a later stage in animal research than that reached by Daubenton. Not that he was more active than Daubenton, but he had started without being handicapped by the problems which confronted the latter. The fact that he concentrated on the development of new individual types and improved breeds particularly interested some French specialists. They did not yet show any actual concern for the improvement of meat. But the value attributed in England to superior rams and ewes seemed to them very significant. They marvelled at the high prices asked for covering. They reported the story of an English farmer, William Story, who sold a three-year old ram for fifteen guineas[3] in 1758. The ram was used at a rate of half a guinea for every ewe served. The names of Robert Gibson[4] and Bakewell[5] were associated with the same ideas. François de la Rochefoucauld visited the latter and his impressions were officially published to the Comité d'Agriculture.[6]

---

[1] *Instruction sur le parcage des bêtes à laine, publiée par ordre du Gouvernement,* 1785. Same work by Lavoisier, 1786.

[2] *Mémoire pour l'amélioration des Bêtes à Laine dans l'Ile de France, par M. le Marquis de G. . . . Paris, 1785. Instruction sur la manière de soigner les bêtes à Laine suivant les principes de M. Daubenton. . . à l'usage des Cultivateurs,* n.d.

[3] Carlier, 'Observations Historiques', *Journal de Physique,* 1784, vol. XXIV, p. 277. [4] *Ibid.*

[5] 'J'ai vu près de Londres, un nourrisseur, le sieur Bekwell en louer cinquante guinées pour le temps du saut seulement', de Guerchy, *Mémoire pour l'amélioration.* [6] Pigeonneau et de Foville, *op. cit.* p. 335.

The beginning of the nineteenth century was to see Bakewell's name as famous in France as it was in his own country; to such an extent was he famous, indeed, that severe criticisms were made of his ideas in order to counter the immoderate influence on the methods of farming fashionable at the time.[1]

On the whole, English influence, both direct and indirect, on the question of breeding was very great. As we shall see further on, it is in this domain that its results are, perhaps, most easily seen, and least confined to the field of theory. Permanent establishments set up for the improvement of livestock and private experiments which opened the way for more extensive developments, prove that the imitation of English agricultural improvements is one of the main causes of the changes which took place in France in the last century. This is largely due to the fact that the increase and improvement of livestock did not originally mean a complete break with the French agrarian tradition.

[1] Especially in Yvart's Report on England (*Mémoire d'Agriculture du Departement de la Seine*, 7 Nov. 1806, vol. x). The quick growth of English animals was criticized by a great many eighteenth-century French travellers, as this meat, overloaded with fat, was said to lose its flavour and consistency when cooking. Yvart takes up the criticisms again and shows the Bakewell products as 'monsters', 'tuns full of oil and fat', and thus strikes his last blow. 'Tandis qu'en France l'anglomanie portait nos agronomes à nous conseiller sérieusement de les imiter en cela, je voyais en Angleterre des milliers de moutons périr de la pourriture. . .' pp. 58, 60, 68. Note the very violent tone of the criticism. Bakewell himself had answered this criticism, see G. E. Fussell, *Agricultural History*, vol. IV, no. 4, 1930, p. 150.

CHAPTER X

CHANGES IN AGRICULTURAL IMPLEMENTS

UNTIL 1840-50 in practice, the agricultural implements used by the whole of the French rural population were almost the same as they had been two or three centuries before. The elaborate sketches of complex machines that can be found in the great *Cours d'Agriculture* of Morogues[1] were undoubtedly no better known in his time than was the drill, for instance, in the time of Tull and Duhamel. In this respect, England was still far more advanced than France.[2] Nevertheless, the second part of the eighteenth century had seen the agronomes lay down the principles of a necessary change in agricultural implements.

The agronomes gave an important place to machinery in their agricultural theories, and we must try to determine what is its position in respect to the whole problem of agriculture at the end of the *ancien régime*, and attempt to estimate how much that position owes to England.

The movement towards innovation of agricultural implements, must be included in the more general one, in existence since the middle of the eighteenth century, for the improvement of technique in the manufacture of instruments and in the mechanical arts. Although engineers were not lacking in France at the beginning of the eighteenth century, it was only in the second part

---

[1] *Cours d'Agriculture*, Plates XL, XLII, XLIII, XCIV, CCCX, etc.

[2] See the important *Recueil de Machines, Instruments et appareils qui servent à l'économie rurale et industrielle*, by Le Blanc, Paris, n.d. (between 1815-30), which shows the considerable number of implements invented and built in England. The French ones do not differ practically from those represented in eighteenth-century books.

147

of the century that application of science to technique made important progress.[1] The movement was illustrated by names like Vaucanson, Montgolfier, Gribeauval and so on. Commonly read and popular books, newspapers, periodicals, dictionaries, encyclopaedias, constantly reproduced articles on and illustrations of, a great variety of inventions. Men like Goyon de la Plombanie, filled several periodicals with all sorts of inventions and projects, in which ingenious ideas can be met together with the most ridiculous ones.[2] Members of the Academy of Sciences concentrated on mechanical problems, and members of the nobility were engineers and inventors.[3]

In this movement prominent place should be given to inventions concerning industry, the connections of which with a similar movement in England have been pointed out sufficiently.[4]

However, in this effervescence of change, any contemporaries whose interest was in agriculture felt that agricultural machinery was making no progress. The *Encyclopédie* noted: 'Cependant la charrue est toujours la même depuis des siècles. . . . Notre charrue n'est pas meilleure que celle des Grecs et des Romains.'[5] Lullin de Châteauvieux wrote:

on reçoit tous les jours avec empressement de nouveaux instruments, de nouvelles machines qui doivent abréger le travail et perfectionner les ouvrages dans les Arts; le premier de tous, l'Agriculture, le plus

[1] For early works on agricultural machinery see: Jacques Besson, *Théâtre des machines*, 1574. Also de Camus, *Traité des forces mouvantes*, Paris, 1722, quoted by F. de Neufchâteau in his report on ploughs (*Mémoires d'Agriculture*, vol. v, p. 98, n. 1). See list of agricultural machines in Rozier's *Nouvelle table des Articles contenus dans les volumes de l'Académie Royale des Sciences de Paris*, Paris, 1775, vol. III.   [2] In *La France agricole et marchande*, Avignon (Paris), 1762.
[3] Counts of Milly and Lauraguais, Marquis of Jouffroy, Count of Caylus, Marquis of Montalembert (from G. Martin, *La Grande Industrie en France*, p. 216).
[4] Germain Martin, *La Grande Industrie en France sous le règne de Louis XV*, Paris, 1900, pp. 185-92, gives an extensive list of English technicians employed in French industry and devotes a chapter to English influence. Also, M. Rouff, in *Les mines de charbon en France, 1744-91*, Paris, 1922, gives instances of English mining machinery directly imported from England, pp. 348-9.
[5] Article: Botanique, vol. II, 1751, p. 343.

admirable, le plus nécessaire et le plus innocent, sera-t-il le seul pour lequel nous ne paraissions prendre aucun intérêt? Sera-t-il continuellement exercé sans principes et offusqué par les préjugés? Cet Art le premier et le plus ancien de tous, est trop soumis aux lois de l'habitude et de l'ignorance; il est temps de venir à son secours.[1]

Must the encyclopaedists be praised, as François de Neufchâteau praises them, for starting such a movement? Must Duhamel be considered as only a disciple of this tendency?[2] It is a point worth discussing. That the agronomical movement has the closest ties with the encyclopaedists is indisputable; but it is hazardous to present its discoveries only as a consequence of the latter school. Thus the question arises of influences on the movement to improve agricultural implements.

All the agronomes agreed that, besides obstacles due to the political and economic order, French agriculture was handicapped by the routine of the peasants. The fact was that perhaps in no field were routine and tradition more unassailable than in that of agricultural tools. But towards the end of the century, genuine cultivators, tenant-farmers for the most part, or enlightened gentlemen farmers, were to be found who actually embarked on a change in their methods and implements. Among these, one can find a certain number of exclusive partisans of sound common-sense, and of a French husbandry which they tried to improve, so to speak from within, while looking for models in technically advanced French regions. A characteristic example of these conservatives is Sarcey de Sutières, appointed by Bertin, Director of an Agricultural School at Annel-les-Bertinval. De Sutières considers, for instance, the plough of Brie the best and most serviceable for French agriculture.[3] Such agronomes thought they were right

---

[1] *Mémoire sur la Pratique du semoir*, Lyon, 1671.

[2] Report of François de Neufchâteau on improvements of ploughs, *Mémoires d'Agriculture*, vol. v. 'Les philosophes français, connus sous le nom d'Encyclo-pédistes, sont incontestablement les premiers qui ont eu l'idée de tourner l'attention des savants vers l'amélioration considérable dont la charrue est susceptible', vol. III, p. 373.   [3] Sarcey de Sutières, *Mémoire sur la charrue de Brie*.

in expecting an improvement in the agriculture of the whole country only through a change in certain French local traditions. One may well ask, however, whether the tendency which caused these agronomes, apparently free from external influence, to desire a modification of agricultural machinery, was not in reality determined as much by the research of those others who studied concrete foreign examples, as by the theory of the 'new agriculture'. The art of watering, for instance, owed something to Italy.[1] The Low Countries and Holland provided the best types of machines for the treatment of textiles, like hemp or flax.[2] The best scythes had to be imported from Germany.[3] But it is England once again which attracted here the attention of the agronomic world. Its agricultural superiority showed itself plainly in the matter of implements which, at the end of the eighteenth century, were the best in the world.

It is interesting to note, however, that at the end of the century, connections between French husbandry and that of other non-English countries, tended to draw closer, and this tendency became more marked at the beginning of the nineteenth century, because of the long break of relations between France and England, and also because of the military expansion of France over Europe.[4] Nevertheless, English influence was so profound that it is noticeable almost everywhere in the field of mechanical research.

Tull had presented new formulas for implements: ploughs,

[1] Reboul, *Discours sur les moyens d'encourager l'Agriculture en Provence*, Aix, 1770.

[2] *Essais de la Société de Dublin, traduits de l'Anglais par M. Thébault*, Paris, 1759.   [3] Morogues, Article: Faulx, *Cours d'Agriculture . . .* , vol. x, p. 396.

[4] This fact is clearly shown in the agricultural reports made at the Agricultural Society of Paris during the Empire. There was a very noticeable reaction (from political motives) against imitation of English methods. The agronomes of this period looked for models in Poland and Prussia. The influence of Thaer in French agricultural theory is significant. Nevertheless A. Young was still considered as the supreme legislator in those questions, and the mission of Parmentier to London in 1802 is a sign that, as soon as the times permitted it, the French agronomic world was still looking towards England.

roller, harrow and drill. His researches were to be enlarged by his French translator, who thus opened to agricultural scientists a new field, which was constantly provided with new material from English discoveries. To archaic conceptions in the matter of implements, the agronomes opposed new and bold views.

The old French plough, whatever its origins, was inadequate and did not respond to the needs of the new method.[1] It often tilled the soil insufficiently and its rusticity forbade any kind of elaborate agriculture. Duhamel tried to modify this state of things and proposed a new plough inspired by Tull's plough with a newly designed ploughshare, sometimes with four coulters.[2] The result was not encouraging. The new plough proved to be much too heavy and provoked much criticism. Lullin de Châteauvieux surpassed Duhamel in his zeal for the improvement of implements, and in this particular field achieved more success. Besides his major role in promoting the drill, he improved Tull's horse-hoe. This light little plough, so handy for cultivation between the rows of crops, Tull's horse-hoe and Châteauvieux's *cultivateur*, remained in the memory of the agronomes longer than the complicated instruments of Duhamel.[3]

As interest in agricultural questions grew, several agronomes began to turn their attention to agricultural engineering. Drawings of ploughs were multiplied in specialized periodicals.[4] The

[1] *Traité*, vol. I, ch. XXII.   [2] *Ibid.* vol. I, ch. XXII, pl. IV, p. 328.
[3] François de Neufchâteau (Report on ploughs, *op. cit.*), opposes Marshall's criticisms of Châteauvieux. In the Article: Cultivateur, the *Encyclopédie Méthodique*, vol. III, p. 692, states 'on nomme ainsi une sorte de petite charrue apportée de l'Amérique Septentrionale par M. St Jean de Crevecœur'. It is the only instance we have found when the *cultivateur* is not attributed to Châteauvieux. Perhaps we must see here a new implement, used in America, which reached France at this time and the shape and use of which were similar to the *cultivateur*. However, the same work, in the article 'Houe', vol. IV, p. 696, definitely states its English origin.
[4] *Journal Oeconomique*, April 1754. Various descriptions of small ploughs, hoes etc. . . in *Journal d'Agriculture, Commerce, Finance*, July 1761; *Mémoires S.R.A.*, 1786; *Feuille du Cultivateur*, no. 67, 23 May 1791.

major importance henceforth attached to implements is echoed in
Le Gentilhomme Cultivateur whose sensitiveness on this point
compels him to issue instructions for the use of the harrow, fear-
ing that careless peasants would only superficially scratch the soil
rather than plough it deeply. These pioneers were opposed by the
conservative agriculturists. Desplaces jeers at Tull's machines and
La Salle de l'Etang gives a piece of advice famous in his day: 'Ne
change point de soc.'[1] Yet, at that time inventions and improvements were proceeding in England with greater practical success
than in France, and these results were known and studied on the
other side of the Channel.

It was the Norfolk plough which was the best known, as it was
partly to its improvements that the French agronomes attributed
the agricultural prosperity of this country. Descriptions of this
plough, still considered by the *Dictionnaire de l'Institut* as one of
the best in England,[2] can be found at the end of the eighteenth
century in the *Feuille du Cultivateur*,[3] in several articles of the
Encyclopédie Méthodique relevant to cultivation, and in the part of
the work devoted to the 'Art aratoire'. The latter work, although
in parts very superficial, may be considered as a sort of précis of
the knowledge of the time on agricultural machinery and reflects
well the new interest in this question. It treats of the Norfolk
plough in great detail,[4] and bases its information on Marshall's
Rural Economy of Norfolk.[5] Arbuthnot, whose important work

---

[1] Legend of the allegorical frontispiece engraved by Cochin at the beginning
of the *Manuel d'Agriculture*. The saying was so famous that in the *Mémoires
d'Agriculture*, vol. III, p. 368, a discussion took place about the exact meaning of
Plinius' 'Sulco vario ne ares' ('Premier rapport fait à la Société sur le développement des charrues').   [2] Article: Charrue, vol. III, p. 404.   [3] 1790, no. 67.

[4] 'Nous avons représenté dans la planche XXXVIII la charrue employée dans
le comté de Norfolk en Angleterre, parce que l'agriculture est très perfectionnée
dans cette province et que cet instrument présente plusieurs avantages que n'ont
pas la plupart des autres charrues.' *Encyclopédie Méthodique*, 'Art aratoire et du
jardinage', 1797, p. 48.

[5] *The rural economy of Norfolk, comprising the management of landed estates and
the present practise of husbandry in the county*, 1787.

on ploughs is published in the *Journal de Physique* (1774),[1] declares this plough valuable because of the height of its wheels which allows a maximum use of the plough with a minimum expense of energy. In France then, plough wheels were generally very low and the agronomes advocated an increase in their height, so that the draughting line be always parallel to the soil.[2] These considerations were studied and used by Despommiers when he built his famous plough (described in his *Art de s'enrichir promptement par l'Agriculture*) which was awarded the prize at a competition on comparative experiments at Châteauneuf sur le Cher, in 1766.[3]

Meanwhile other models of ploughs continued to arrive from England. Arbuthnot's article was widely broadcast. His researches on the improvement of the ploughshare were continued, thus establishing a link with Duhamel's research on the same subject. Arbuthnot had endeavoured to discover the best shape for the mould-board and whether, by giving the share a semi-cycloidal or semi-elliptical shape, its wear, or the rubbing and agglutination of the earth, might be prevented. The question of rubbing preoccupied agricultural engineers. They recommended the fitting of the blade of the plough with an iron plate which allowed it to slide better, with less adherence—an invention which came from England.[4] Such ploughs were called 'charrue anglaise sans roues'[5] and were used at Liancourt three or four years before the Revolution.[6] At the close of this period, English ploughs were

---

[1] 'Mémoire de M. J. Arbuthnot, écuyer Anglais, Membre de la Société Royale . . . sur les principes et construction de sa charrue', *Journal de Physique*, vol. IV, pp. 284-97.

[2] One of the conclusions of Article: Charrue, *Dictionnaire de l'Institut*.

[3] Article: Charrue, *Dictionnaire de l'Institut*, vol. III, p. 404. *B.P.E.* 1788, vol. II, p. 1 ff.　　　　　　[4] Rozier, Article: Charrue, *Cours d'Agriculture*.

[5] *B.P.E.* 1789, vol. II, p. 1; 1791, vol. II, p. 19.

[6] *Mémoires d'Agriculture*, vol. III, p. 425. In 1814 the Duke writes, 'Je fais usage chez moi de la charrue anglaise de Suffolk, comme sous le nom de swimming plough ou charrue nageante, et je m'en trouve bien.'

increasingly favoured: 'dans ces derniers temps, les Français ont oublié leurs propres succès en ce genre et leur enthousiasme s'est tourné presque exclusivement vers les charrues anglaises.'[1] Mortimer's *Treatise of Husbandry* gives popularity to the Colchester ploughs, and those of Lincoln, Sussex and Caxton. An article by Ellis, 'Fermier à Gaddensden, Hertfordshire', in the *Gentleman's Magazine* of February 1770, translated in an article of the *Encyclopédie* (Supplement of 1776), describes a double plough. At the Royal Society of Agriculture, two Picard farmers presented a two-coultered and two-shared plough, but stated that the first credit for its invention should be given to the English, and they quoted a description of ploughs by Sharp.[2] These specimens were completed with descriptions of Mr Cook's iron plough, and Lord Sommerville's 'two-furrow swing plough'.[3] The period ended with the introduction into France of Arthur Young's plough, famous after its victory in a competition in Suffolk, and also used at Liancourt 'avec un succès toujours soutenu . . . et par conséquent peut l'être partout'.[4]

Of all these ploughs, it is true to say that none survived in practical agriculture in France; for the changes and improvements introduced very seldom proved simplifications. As a matter of fact, the eighteenth century invented in this domain machinery that was too complicated, so that the peasants—persuaded only by simple things, as the agronomes said—were diffident about it. Moreover, the anglomanes had to admit that the English themselves did not always agree about the merits of their implements, and it was hard to recommend a plough which was not unanimously accepted even in its own country. Nevertheless, it is less the actual

---

[1] *Mémoires d'Agriculture*, vol. III, p. 368, 'Premier rapport fait à la Société sur le perfectionnement des charrues.'
[2] *Mémoires de la Société Royale d'Agriculture*, 1785.
[3] *Ibid.* vol. III, p. 431. A certain Mr Johne's double furrowing plough had already been used in 1766 by farmers at Pomponne, Villemard, Noisy-le-Grand.
[4] Article: Charrue, *Dictionnaire de l'Institut*, p. 405.

use of these ploughs which counts, than the activity in research they stimulated and which was later to be co-ordinated by Mathieu de Dombasles, successor of the eighteenth-century agronomes.

Ploughs, however, were not the only traditional implement to undergo change, thanks to England's example. In France, it seemed, both the simplest and the most complex implements had been the same for centuries. In England, on the contrary, the eighteenth century saw a prodigious activity in innovations. Baert, describing the general perfection of implements in England, wrote, 'on invente tous les jours en Angleterre de nouveaux instruments d'agriculture dont le succès n'est pas toujours constant. Aussi y sont-ils en général très multipliés, très variés'.[1]

Among these improved implements the agronomes pointed out some variations in the simplest ones, which allowed a better use of them. We find, for instance, that the spade was susceptible of modifications. According to their different uses, the spade of Lincolnshire for cutting peat, the spade of Essex with a wide iron, the spade of Herefordshire with a sharp crescent-shaped iron, were all recommended in France.[2] Experiments had already been made on this point, but on a very limited scale.[3] The state of English agriculture proved that the smallest details had their importance. The French discovered also that it was 'time to make improvements in the harrow' and broadcast Dr Home's teaching, presented to the public his harrow with sharp teeth and his double harrow.[4] The cleaning up of heaths was to be facilitated by means of the 'dog'—a kind of hook for the extraction of roots.[5] After the failure of the drill the *Feuille du Cultivateur* tried to

[1] *Tableau de la Grande Bretagne*, Paris, An. 8, vol. III, p. 263, n. 1.

[2] Article: Bêche, *Encyclopédie Méthodique* ('Art aratoire'), p. 21.

[3] *Journal Oeconomique*, 1751, Article on a rotative spade invented by the Count de Crequi-Frohans, in Picardy.

[4] Article: Herse, *Encyclopédie Méthodique* ('Art aratoire'), p. 115. Also 'Nouvelles herses propres aux usages de cet instrument suivant la qualité et l'état des terres, par M. Home', *B.P.E.* 1788, vol. II, pp. 5 ff.; *ibid.* 1787, vol. I, pp. 1 ff.

[5] Article: Dog, *Encyclopédie Méthodique* ('Art aratoire'), p. 80.

advocate the dibble, because it was much used in England,[1] and the hand-hoe 'dont l'usage est fort commun en Angleterre'.[2] All these details show how very extensive was the interest aroused by English implements, which were considered as representative of a high technical level.[3]

But the curiosity reached its climax over the new English machines, symbols of a mechanized agriculture, which were to provoke enthusiasm, as well as stupefaction and disillusionment, in France: the drills, threshing machines, harvesting machines, and mowers aroused the curiosity of travellers. And just as the export of sheep was strictly prohibited in England, so there were restrictions about the divulgence of mechanical secrets. The Abbé Morellet gives an interesting account of his stay at Lord Shelburne's in Wiltshire, which shows how techniques were sometimes transferred from one country to another. Marvelling at the agricultural implements he found there, he asked for

quelques modèles, non pas de machines dont les Anglais sont jaloux à l'excès, modèles qu'on n'eut pu me donner et que je n'eusse pu emporter avec sûreté; mais des dessins et autre objets non moins utiles, comme un dessin et une explication de mettre le foin en meule, de le couvrir avec un toit mobile qui descend à mesure que la meule se consume; un couteau à couper le foin, les diverses mesures de contenance et de longueur et les poids étalonnés; de nombreux échantillons etc. . . toutes choses que j'ai rapportées pour le gouvernement.[4]

Similarly, Lazowski in 1788 comments on the existence in England of several kinds of agricultural implements still hardly known in

---

[1] In nos. 74 and 80.

[2] Article: House, *Encyclopédie Méthodique*, vol. IV, p. 696.

[3] It would, however, be exaggerating to state that all changes were determined by England. Improvements of the sickle and of the scythe came from Flanders or Germany. In 1786 the Parlement of Douai still forbade the use of the scythe (which left very short stubble and so deprived the poor of one of their resources). It is interesting to note that opposition to the scythe was also strong and lasting in England itself (Moffit, *op. cit.* p. 4).

[4] *Mémoires*, Paris, 1822, vol. I, p. 213. Similarly De Mante, *Traité des prairies artificielles* . . . , states that in 1774 he had drills and horse-hoes sent from England.

France. Among them he quotes a 'rabot propre à raboter les terres et unir les terrains', a ventilator, a threshing machine, a turnip-cutter, a chaff-cutter, a combine which ploughed severall furrows, sowed and harrowed at the same time, a winnowing-machine and so on. Lazowski finds these inventions so interesting for French farming that he recommends making a collection of these machines for exhibition.[1] The proposal was partly realized and was a prelude to the creation of the Conservatoire des Arts et Métiers, some years later.

Among the new inventions, the drill is certainly the one which in this period had the greatest vogue and the longest life.[2] This implement was considered indispensable for the new husbandry and Duhamel wrote in the *Eléments* that the whole system could be reduced solely to the use of the drill.[3] A picturesque account was given of the machine: 'L'ouvrage de M. Duhamel réveilla l'attention de tous les cultivateurs et grands propriétaires. Chacun voulut avoir un semoir et perfectionner celui de M. Tull . . . alors on offrit à la curiosité publique les semoirs à tambour, les semoirs à cylindres, les semoirs à palettes.'[4] MM. de Châteauvieux, de Montéfui, Diancourt, Thomé, Blanchet, de Villiers added simplicity to the original model.[5] But the machine was still far from perfect. The example of England, where the attempts of Mr Ellis, Dr Hunter and Mr Rundell had failed, gave birth to a 'seminomania',[6] proving, however, that the new invention was impracticable. 'Aujourd'hui tous les semoirs sont relégués sous le hangar et on ne s'en sert plus.'[7]

[1] Pigeonneau et de Foville, *op. cit.* p. 351.

[2] See the complete list of drill inventors, with technically interesting details on French drills in Russel H. Anderson, *Agricultural History*, 1935, vol. x, no. 4, pp. 172-3.   [3] *Eléments*, vol. I, p. 483.

[4] Article: Semoir, *Dictionnaire de l'Institut*, vol. XI, p. 458.

[5] *Traité*, vol. VI, ch. IV, also quotes MM. Tulle of Avignon, Chevalier de Voussi, de la Tasse, de la Levrie.

[6] In 1787 Cook was in Paris waiting to show his last drill to the Duc d'Orléans, in A. Young, *Travels*, vol. I, p. 184.

[7] Article: Semoire, *Dictionnaire de l'Institut*.

The very form of the drill prevented it from being popular. It was too complicated and did not work very well. The great problem was an equal distribution of the seeds, and the avoidance of ground between the cylinders or paddles. Thus the last famous drill in France was the Abbé Soumille's built in 1763, and praised by Rozier. It was a mere wheelbarrow out of which the seeds were uniformly sifted into a furrow traced by a little coulter in front of the machine. Although simple, it was not more used than the preceding ones.[1] The drill met throughout the same opposition as the new husbandry, of which it was a symbol. It had nevertheless its apostles, like Lullin de Châteauvieux, who displayed a prodigious activity in his efforts to extend its use. He advocated it in his own district, sent many specimens of it to remote parts of France. In his *Mémoire sur la Pratique du Semoir*, he confirmed all the advantages of the drill by a certificate from M. de la Michodière, Intendant of Lyons, who attended the experiments. In 1762 the Agricultural Society of Alençon had also published a paper saying that the use of the drill saved four-fifths of the seed employed.[2]

In spite of these wonderful results, success did not come. Sowing at the end of the *ancien régime* was still done by hand, and the drill remained a strange symbol of indoor agriculturalists.[3]

It was to be the same with a most ingenious invention related to the problem of the preservation of grains: Duhamel's granary. This question was of particular importance in the restrictive

[1] Soumille, *L'usage du semoir*, 1755, enlarges and simplifies Duhamel's implement. The Abbé's drill being stronger and simpler was always considered better than others. The last example of a complicated drill exists in Rey de Planazu's *Oeuvres d'Agriculture*, 1786. The idea of a drill, however, never entirely died. The *Mémoires d'Agriculture* still advised its use in 1803 and said 'qu'il y a dans la seule ville de Londres, quatre fabriques de semoirs'.

[2] Instead of using 300 lbs of seeds according to the common practice, 60 lb only were used with Duhamel's drill.

[3] About the use of the drill in England at the same time, see, A. S. Haslam, *The Biography of Arthur Young*, Rugby, 1930, pp. 45-6.

legislation about grains. Harvested grains were stocked either in the granaries of the peasant producer, or in huge stores, designed to supply a district or a fortress, or constituting stocks *d'abondance* for the years of dearth. Several accidents threatened grain thus stored, and it happened that whole crops were lost because of ignorance about methods of preservation. This problem was not as acute in England. Movement of grain was different from what it was in France,[1] and in consequence, the question of granaries was not of such importance. Again, the better quality of English products made them less liable to deterioration (the study of grain diseases was far less developed in England than in France) so that they were easier to preserve. In France the problem was to prevent grain from rotting or being eaten by insects. Curious instances are found in treatises of the time, stating that wheat had been preserved for a very long time, because in one way or another it was sheltered from the surrounding air and moisture.[2] Duhamel once more applied himself to the problem and invented a new machine to solve it. He discovered that constant renewal of air around the grain, would prevent fermentation and multiplication of weevils. The whole question reduced itself to circulation of air through a huge mass of grain. The question of ventilation was then being studied in Sweden and in England. Duhamel intended at first to use a system worked out by Martin Triewald, a Swedish engineer, which permitted ventilation of ships' holds;[3] but it was feared that rats might tear the leather of the bellows. 'Il lui vint d'autres idées auxquelles il avait pensé à s'arrêter: enfin il était dans l'embarras du choix quand M. Halles lui fit parvenir un exemplaire de son ouvrage intitulé, "le Ventilateur". . . . M. Duhamel ne tarda pas à adapter à son grenier, le

---

[1] L. W. Moffit, *op. cit.* pp. 74-5.

[2] Instance of the Metz citadel was given where corn stocked in 1709 was rediscovered intact 20 years later, having been preserved by a crust of conglomerated grains with dust, in Bomare, *op. cit.* Article: Blé, vol. I, p. 424.

[3] Article: Conservation des grains, *Encyclopédie Méthodique*, vol. III, p. 464.

soufflet de M. Halles.'[1] In this ventilator, the leather bellows were replaced by a wooden piston. Duhamel had thus solved his problem thanks to his relations with English scientists. The results of the new granary were rather remarkable, according to the *Traité de la Conservation des grains*, and the writings of the agronomes.[2] But the care with which the new machine had to be built, the quality of materials necessary and the relative complexity of its mechanism, in one word the general cost of the work, prevented its use on a large scale. Nevertheless, a solution had been found to the problem of the storage of grain.

At the same time Duhamel's personal researches, as well as reports from England, were also to enlarge the results of experiments performed since the beginning of the century on complex machinery such as the threshing-machine. It is difficult to say who was the first to attempt its construction. It had actually appeared in France in 1737, perhaps earlier.[3] A general Commissaire in the Royal Studs of Provence, Meffrein, presented to the Academy of Sciences a threshing-machine which did the work of six vigorous men. Before Meffrein, two engineers, Mossadigni and Duquet, had invented a cranked handle which set in motion several flails at the same time.[4] These inventions seem to have had no effect in the agricultural world. Thus, we must attach more importance to similar work carried out in England by Ewert of Swillington, and still more by Andrew Meikle, twenty years or so later.[5] These machines were certainly known in France, but they were said to crush the straw and this, for a time, prevented any fur-

---

[1] *Idem.* About Hale's Ventilator see Berman, *History and Art of warming and ventilating rooms and buildings*, London, 1845, vol. II.

[2] Also *Supplément au Traité de la Conservation des grains*, Paris, 1756 and *Journal Oeconomique*, October, 1752, pp. 57 ff., 83.

[3] The list of articles of the Academy of Sciences, made by the Abbé Rozier, gives instances of such machines in 1716, 1722 and finally, the Meffrein one in 1737.

[4] *Mémoires de l'Académie Royale des Sciences*, 1722.

[5] The *Nouvelle Maison Rustique*, edition of 1749, had already advertised a new English threshing-machine (vol. I, p. 651).

ther interest being taken in them. It was considered very important that straw should always be sold in unaltered condition. Crushed straw or straw in bulk was not accepted in the Paris market.

But the invention was being improved. It may be supposed that English threshing-machines influenced French inventions. Morogues states that Meikle's machine, for instance, went from Scotland to England, from there to Sweden, then to France.[1] At the end of the period, Rey de Planazu invented a threshing-machine which had some relationship to that of Meikle.[2] This invention was not more successful than the previous ones. However, the idea of a threshing-machine survived. At the end of the period the *Encyclopédie Méthodique* was still advocating the use of a machine for threshing as 'tout ce qui tend à abréger la main d'oeuvre doit être précieux à la société'.[3] Besides these intricate new inventions, simpler ones were also known and sometimes used. Such were the chaff-cutter and the turnip-cutter. The chaff-cutter, some models of which were also German, was said to have been improved by a Mr Smith who invented one with two knives.[4] Planazu boasted of having improved the English chaff-cutter 'compliqué et facile à se déranger'. Mr Edgill's root-cutter was analysed together with a mill for grinding oats and other grains for feeding horses, 'much used in England'.[5] Although the actual employment of these implements was very limited[6] it is nevertheless interesting to see how mechanized agriculture, inspired by models from England, penetrated into eighteenth-century France.

---

[1] Article: Battage, *Cours d'Agriculture*, vol. III, p. 203.

[2] Calonne, *op. cit.* p. 96. See plan of the machine in Planazu, *Oeuvres d'agriculture*, no. 12, 1786.

[3] Article: Batteur en Grange, *Encyclopédie Méthodique* ('Art aratoire'), p. 15.

[4] *B.P.E.* 1787, vol. I, p. 84.   [5] *Ibid.* p. 88.

[6] They were, however, introduced in certain districts by the great landlords. M. Engalric near Narbonne used two chaff-cutters in 1812, 'dont une a l'imitation du modèle Anglais', in 'Tableau des améliorations introduites depuis environ cinquante ans dans l'Economie rurale de l'arrondissement de Narbonne', *Mémoires d'Agriculture*, vol. xv, p. 243.

There is also a streak of English influence in another field of research, which does not, properly speaking, concern machinery, but is related to the fitting up of farms or country estates.

Buildings, for instance, as they traditionally existed, were criticized. The agronomes found the different outhouses on a farm far too numerous, entailing unnecessary repairs and expensive maintenance, in addition to perpetuating certain unwholesome traditions. The example given by England showed a simplification of this elaborate set of buildings, as the English peasant tended to get rid of his corn as soon as possible, and left his animals and straw outside all the year round, with proper but light shelters. But rural building being, more than any other side of agricultural life, adapted to the soil and climatic conditions in which it is placed, seems less susceptible of change. Thus, the agronomes never cared to advocate the conversion of French farms into English homes.[1] They did, however, present English results, and it is possible to find in certain regions the beginning of a movement for a better use of the farm buildings.[2]

In other parts of the farm or the estate, English discoveries or tested means helped in solving many problems.[3] For instance, the technique of hedge-growing for enclosures was considered as very English. In the important articles devoted to this question in the *Encyclopédie Méthodique* and elsewhere, reports of English

---

[1] However, this was a subject of study for many of them. The emphasis was finally given to this tendency with Lasteyrie's important book, *Traité des Constructions rurales . . . ouvrage publié par le Bureau d'Agriculture de Londres et traduit de l'anglais avec des additions*, Paris, An. x, 1802.

[2] *Mémoires d'Agriculture*, 1812, mentions 'Lorsque les communications avec l'Angleterre et l'intérieur de la France devinrent plus fréquentes, les premières améliorations se portèrent sur les instructions des maisons, des granges, des colombiers, des écuries et autres bâtiments', vol. xv, p. 360. This happened around 1770.

[3] In connection with improvements in vegetable cultivation, Morogues (*op. cit.* Article: Cultures forcées) writes an interesting account of greenhouse building at the end of the eighteenth century. He relates it to the same movement in England, vol. VIII, p. 148.

methods are constantly met with: 'Nous aurons plus d'une fois l'occasion de citer dans cet article, l'exemple des Anglais, comme nous avons également profité des instructions que plusieurs de leurs agronomes ont publiées sur la formation des différentes espèces de clôtures.'[1] The *Feuille du Cultivateur* proposed to its readers a 'clôture peu dispendieuse imaginée par un cultivateur Ecossais', a wall of stones topped by a bush hedge.[2] The *Gentilhomme Cultivateur* and the *Voyage Agronomique* rather tend to advocate English enclosures with fences and turf.[3] The question of the nature of the hedge itself often raised controversies, and the standard books on the question quoted the English solution in order to settle the point.

The fitting up of fields or pastures included the problem of drainage, itself connected with the intensive movement for the reclamation of lands in the second part of the century.[4] This was another question to which English technique had supplied the answer. Gilbert, noticing that meadows north of Arras were constantly flooded, complained that the system, 'importé d'Angleterre par Mylord Ogilvy' was not even applied.[5] Rozier, on the same question of drainage, proposes a local custom, 'connue en Angleterre',[6] a sort of underground drain-pipe system. But some time later it was found that the new technique was drying up the lands, and that what was beneficial for corn lands was detrimental to pastures: 'Il ne faut donc pas que la manie de l'imitation nous porte trop loin.'[7]

There are finally other sorts of agricultural implements of a lesser importance, which were nevertheless advertised in France,

[1] Article: Clôture, *Encyclopédie Méthodique*, p. 298.
[2] *Ibid.* p. 304. [3] *Ibid.* pp. 305-6.
[4] On the increase of mill ponds and their bad consequences for natural pastures, see Carlier, 'Mémoire', *Journal de Physique*, 1784, vol. II, p. 101; De Pradt, *op. cit.* vol. I, p. 142.
[5] Gilbert, *Mémoire à l'Académie d'Amiens*, 1787, in Calonne, *op. cit.* p. 124.
[6] Article: Dessèchements, *Dictionnaire de l'Institut*, vol. IV, p. 509. [7] *Ibid.*

as they were in direct relation with the new system of husbandry. Among these are the different kinds of sheep folds, the complexity of which increases the regularity of manuring.[1] In connection with the question of manures and fertilizers, the new interest shown in marling inaugurated a movement of research on soils. This led to the importation into France of instruments for exploration underground, and different English models of bores were either purchased or imported.[2] Valmont de Bomare pointed out the opportunity 'représenter au gouvernement combien il serait utile d'avoir dans chaque district de ce Royaume une grande tarière banale pour sonder la terre. . . . La dépense d'une telle sonde est peu considérable et l'utilité en serait très grande.'[3] We find in Patullo, in Turbilly, and in Planazu that the necessity of analysing the soil by means of instruments had been fully realized, in the light of the English achievements in mixing soils.

The equipment of the countryside with roads and canals was also part of England's agricultural fame. Although roads (mostly highways) were generally much improved in France at the end of the *ancien régime*,[4] a great deal remained to be done as regards smaller country roads. These were generally damaged by the narrow wheels of overloaded carts and the problem was so pressing that Trudaine, head of the Service des Ponts et Chaussées, was much concerned about it.[5] English legislation had imposed a certain width for cart-wheels in order to prevent such inconvenient occurrences.[6] Trudaine had collected in 1754, 1755, 1757 and 1763

[1] 'C'est la méthode d'Angleterre et celle du pays de Caux', in *Instructions sur le parcage.*   [2] Pigeonneau et de Foville, *op. cit.* p. 315.

[3] Article: Marne, *Dictionnaire d'Histoire Naturelle*, vol. III, p. 697.

[4] E. J. M. Vignon, *Etudes historiques sur l'administration des voies publiques en France aux 17e et 18e siècles*, Paris, 1862, vols. II, III.

[5] 'Mémoire sur la nécessité de faire adopter, sans exception, l'usage des roues à larges jantes pour les transports de toute espèce, par M. Savoye-Rollin, préfet du département de la Seine Inférieure', *Mémoires d'Agriculture*, vol. x, pp. 162 ff.

[6] Montague-Fordham, *A Short History of English Rural Life*, London, 1918, p. 120, Appendix, p. 173.

all the bills of the English Parliament on this matter and proposed them as a model. But as usual, the government did not take any definite step. They gave the greatest publicity to the new idea and distributed wheels with wider fellies, but in vain. The question was later taken up again by Liancourt before the Provincial Assembly of Soissons.[1] The end of the eighteenth century was also the time when canals in England had begun to be improved and new ones were extensively built.[2] Although there was a standard French book on the question[3] the fresh example set by England contained new material disclosed in Hugh Hensdall's book which, according to the *Journal de Physique* could 'suppléer à ce qui manque dans le grand ouvrage de M. de Lalande sur les canaux navigables'.[4]

We see that no detail capable of improving French rural life was neglected by the agronomes; England was showing the results of a better application of mechanics to agriculture and was offering the picture of a countryside where life was made easier for the farmer. Thus they advocated imitation of England. In the question of machinery France was, more than in any other field, dependent on England, as its industry was less developed in that particular direction and was also less perfect.[5] It was only from England that complex and useful machinery could be obtained. The agronomes followed the example of the mine owners or great industrialists in this matter. But the creation of a modern industry had fewer obstacles to overcome than the modernization of a traditional agriculture which, in addition, did not even possess the financial means for buying the necessary implements for such a

[1] F. Dreyfus, *La Rochefoucauld-Liancourt*, Paris, 1903, p. 48.

[2] Ernle, *op. cit.* pp. 277-8. Moffit, *op. cit.* pp. 138-9.

[3] Lalande, *Des Canaux de Navigation*, 1778.

[4] 'A Plan of the Navigable Canal. Plan des canaux navigables qu'on a fait et qu'on fait maintenant en Angleterre.' Londres, chez Lowndes, *Journal de Physique*, 1780.

[5] These are the conclusions of Chaptal in his famous book, *De L'Industrie Française*', Paris, 1819, vol. II, pp. 31, 32, *passim*.

radical change. That is why the French agronomes were often called Utopians, and their research was judged if not ridiculous, at least unsuccessful. Half a century later, owing to the freeing of the land and the industrial Revolution in conjunction, English methods and English implements were to be studied again, and this time with success. But those who were the pioneers of the transformation in the eighteenth century were by then forgotten.

CHAPTER XI

BEGINNINGS OF AGRICULTURAL CHEMISTRY[1]

THE new husbandry appeared as a well co-ordinated system which, based on observation and experiments, was prepared to answer any objection raised against its internal structure. A question, however, remained to be answered. It seemed obvious that some regions, owing to the quality of their soil, or the peculiarities of the climate, could not benefit by the new discoveries. The agronomes had already had to counter the hostility shown, to a certain extent, towards things English. And the criticism was easy to make that geographical conditions not being the same in England and in France, the success of a system in the one country did not necessarily spell its success in the other.[2] The adaptation of the new husbandry to the soil was one of Duhamel's main concerns, and his insistence on the Tullian theory of repeated ploughings was in fact meant to show that manures should not be considered as the only means of enriching the soil, and that there were other means of bringing out all the nutritive substances contained in it. In other words, that few soils only would not, by a certain process, give the plants their necessary food: 'pour

---

[1] The question of organic chemistry and its relations with certain special techniques has not been dealt with. However interesting a study of the act of baking, of wine making, of sugar refining, etc., would be, it was felt that this was a problem beyond the scope and without the limits of this work. Information on the particular points mentioned above can be found in the works of Parmentier, Macquer, Guyton de Morveau in particular, and all the works of eighteenth-century French chemists in general.

[2] *B.P.E.* 1782, vol. II. Important introduction which analyses the exaggerations of the new tendencies; 'on vante trop les nouvelles cultures sans savoir assez bien où elles s'adaptent.'

augmenter la fertilité des terres, il ne suffit pas de les pourvoir de la substance qui doit nourrir les plantes; il faut de plus les disposer de façon que les plantes puissent recueillir avec leurs racines ces mêmes sucs que presque toutes les terres contiennent abondamment.'[1]

For the first time in France, Duhamel was setting, and setting in precise terms, the principal problem of fertilizing ground on which a new system of husbandry was to be applied.

Up to that time, the solution proposed by the old agricultural authors had been rudimentary. Fallowing and manuring were its major factors, making of the agricultural process a vicious circle. The use of different fertilizers according to the different soils, however, had led to a classification of sorts, which varied according to the authors. The discrimination that led to this classification lacked coherence; we see Liger, for instance, establishing his categories on physical principles ('terres fortes'), or descriptive ones ('terres pierreuses'), or chemical ones ('terre de craie'), or even temporal ones ('terres novales'). Each of these had to be given a different kind of manure, chiefly dungs, the qualities of which were generally well understood, and sometimes vegetal fertilizers of different kinds;[2] marling and the mixing of soils was also advised,[3] although to a lesser extent. Yet, in fact, no coherent system of fertilizing had ever been set up.[4] Tradition and a vague empiricism prevailed.

Réaumur, some time before 1750, had been interested in 'chymie agricole'[5] but the real impulse was to come from the experiments carried out in England. The discovery of English agricultural methods around 1750, the expansion of Tull's principles in the

---

[1] *Traité*, vol. I, ch. VI, p. 51.

[2] Liger, 'Premiers Labours', *Oeconomie Générale*, vol. I, p. 231.

[3] *Ibid.* vol. I, p. 225. 'Manière d'améliorer les terres par les terres mêmes, par Goyon de la Plombanie', *Nouvelliste Oeconomique*, 1754, vol. II, p. 39.

[4] See the demonstration of this fact in Liebig, *Traité de chimie organique*, Paris, 1840, p. 113.　　　　　　　[5] Weulersse, *op. cit.* vol. I, pp. 347, 348.

Traité de la Culture des Terres and of Duhamel's personal *Observations Botanico-Météorologiques*,[1] and the elaboration of the doctrine of crop rotation, initiated in France the study of agricultural chemistry. Strictly speaking, the actual scientific value of the researches made is very poor, since no chemical explanation was in fact given for the phenomena observed. None the less, they deserve passing mention for the interest they aroused in a matter hitherto considered as only secondary, and for their revelation of many a fact, finally established shortly after the end of our period.

It had been agreed that manures were as important as ploughings in agricultural operations. The English, in fact, far from rejecting their use, knew much more about them than the French. The science of manuring seemed more developed in England than in France.[2] The *Encyclopédie* quoted in detail the manuring system of Norfolk. The strangeness of certain materials employed, besides dungs, was the subject of many articles in the *Journal Oeconomique*.[3] But the interesting aspect of the movement is the publicity given in France to Lord Townshend's action in reviving marling and the mixing of soils.[4] Patullo, advertising his country, wrote: 'Il est maintenant généralement connu en Angleterre . . . qu'il y a très peu de terres qui ne contiennent dans leur propre sein, des engrains propres à en améliorer la surface.'[5] Analysis of soils was in fact rapidly improving in England. In the *Eléments du Commerce*, Forbonnais gave a classification of soils in accordance with English principles and even kept in the French text the English names of soils: loams, steel-marle, chalky-lands, etc.[6] Such was the importance attached to English discoveries, that these can be found in common use in many an agricultural book of the time. But the first to emphasize the importance of 'la

---

[1] See below, p. 173, n. 3.
[2] Moffit, *op. cit.* pp. 22, 23; *L'Ami des Hommes*, vol. v, pp. 163-85.
[3] Especially sea-sands, sea-shells, sea-water, etc. . . (1751-5).
[4] In the Norfolk Letters quoted in ch. II, see Bomare, *op. cit.* Article: Marne.
[5] *Op. cit.* p. 13.   [6] *Eléments du Commerce*, p. 217.

diagnostique des terres' which, according to Patullo, was still very little advanced,[1] was Francis Home, whose book, *The principles of agriculture and vegetation* was translated in 1761.[2] The French agriculturists were shown for the first time, by a Scottish doctor, that a good agricultural practice 'dépend de principes que sa pratique seule ne peut apprendre'.[3] Home was setting chemical foundations to Agriculture; 'sans la connaissance de cette dernière science [chemistry] il n'est pas possible d'établir les vrais principes de l'agriculture. Or la Science de la chymie ne faisant, pour ainsi dire, que de naître . . . on ne s'était presque point aperçu de la liaison que l'agriculture a naturellement avec elle.'[4] Besides giving a reasoned distinction of different soils, his endeavour was to 'detect the constituent parts of these different soils', and 'to take a view of those qualities belonging to them, by which they are capable of operating on the soil and producing certain effects on the vegetation of plants'. Home's work is still analytical rather than demonstrative. However, his contribution is important in that he lays the emphasis on questions which the limited chemical knowledge of his time could not solve, but which demanded further research.

Thus he supported, in the French agricultural world, the efforts already undertaken by Duhamel and mainly Patullo for directing experiments towards a better understanding of the differences of soils and the manures they required.[5] Wolters gives the year 1760 as the starting point in France of agricultural chemistry.[6] It is true that after this date observations and accounts of experiments in this field multiplied. More importance was being attached to

[1] *Op. cit.* p. 23.

[2] *Les Principes de l'Agriculture et de la Végétation, ouvrage traduit de l'Anglais de Mr François Home*, Paris, 1761.

[3] Wolters, *op. cit.* p. 20.

[4] Home, *op. cit.* p. 4.

[5] Paring and the burning of turf for reclaiming barren lands was also advocated (Duhamel and Turbilly) with frequent allusions to this method in England.

[6] Patullo, *op. cit.* p. 23.

the quality of animal manures that Patullo had criticized. The Royal Society of Agriculture proposed in 1765 the following subject for its prize: 'Le meilleur Travail sur la Qualité et sur l'Emploi des Engrais qui conviennent aux Terres, principalement aux Terres à Blé, relativement à leur Qualité.' In 1767 M. de Massac published a 'Mémoire sur la qualité et sur l'emploi des engrais', a work which the Society of Agriculture of Berne greatly encouraged.[1] In 1778 the Society of Limoges asked which were the fertilizers to use for 'le haut Limousin et à remplacer la marne qui y manque.'[2] The favour attached in England to marling, and its success, particularly in Norfolk, made the deepest impression in the agronomic world. Valmont de Bomare refers his readers to the *Corps complet d'Agriculture d'Angleterre*[3] on this subject. Roland says that 'La pratique de marner les terres est aussi très répandue en Angleterre',[4] while an article of 1783 proposes a means of making artificial marl with clay and chalk 'selon un Académicien de la Société Royale de Londres'.[5]

Towards the end of the period various combinations of different kinds of minerals were advertised and experimented with. The technique of liming was particularly developed. Rozier, in an extensive article, describes the method used by 'M. Mill, célèbre agriculteur anglais, d'après les instructions remises à la Société d'Edinburgh par M. Lummis'.[6]

In 1786 a 'particular manner of using lime as fertilizer' by a Mr MacLure of Shawrod [*sic*] was also proposed.[7] Mr Andrews' method, which mixed animal manures and lime awoke the interest of Planazu.[8]

Besides lime, different methods of using ashes, peat and coal,

[1] Article: Massac, *Biographie*, Michaud.
[2] *Journal de Physique*, 1778, vol. I, p. 194.
[3] Hale's, *Compleat Body of Husbandry*, London, 1758-9.
[4] Roland de la Platière, 'Mémoire sur l'éducation des troupeaux. . .', *Journal de Physique*, 1779.   [5] *B.P.E.* 1782, p. 347.
[6] Article: Chaux, *Cours d'Agriculture*, vol. II, p. 262.
[7] *B.P.E.* 1786, vol. I, p. 19.   [8] *B.P.E.* 1787, vol. I, p. 5.

and composts[1] came from England. Adaptation of certain fertilizers to certain plants was accomplished with greater precision. Among the numerous experiments performed in that field, fertilizing clover with plaster or gypsum proved one of the most successful.[2] At the very end of the century this practice was being scientifically studied by people like Scanegatty, professor of physics in Rouen, and Faujas de St Fond.[3]

Yet even then, and in spite of Arthur Young's assertion that the use of fertilizers seemed well understood in France, the agronomes did not seem satisfied with it. Tessier, in his article 'Fumiers', still criticized the French way of using manures and based his suggestions on Kirwan's *Traité des engrais*. The Comité d'Agriculture still deplored the fact that 'La science des engrais n'existe pas encore. On en ignore les premiers éléments. On ne sait pas ce que c'est que du fumier, comment et pourquoi il fertilise la terre. Quels sont ceux qui conviennent le mieux aux différentes natures du terrain? On n'a sur tous ces objets que des routines peu éclairées.'[4] And Baert still marvelled, on his visit to England, at the number and variety of products used as manures.[5]

Another way of fertilizing the soil, different from that of using organic or mineral elements, was also at least suspected, if not thoroughly studied, in the second half of the eighteenth century. This was the beneficial effect of certain crops. The fact had been observed in close connection with the application of the doctrine of rotation. Although the reasons for the phenomenon were unknown (Duhamel's embarrassed explanations are a typical in-

[1] *B.P.E.* 1790, vol. I, p. 20.
[2] 'Manière dont on se sert du plâtre dans quelques cantons du Dauphiné pour les prairies artificielles', *B.P.E.* 1782, p. 83. See a report on the first attempts made with gypsum in *Journal de Physique*, 1774, vol. IV, p. 18.
[3] Arthur Young, *Travels in France*, ed. by H. Sée, vol. III, pp. 1237, 1238.
[4] G. de la Fournière, 'Les Comités d'Agriculture de 1760 et de 1784', *Bulletin du Comité des Travaux historiques et scientifiques* (Section des sciences économiques et sociales), 1909, p. 117.
[5] *Tableau de la Grande Bretagne*, vol. III, p. 247, n. 1.

stance of it)[1] the fact was discovered that some 'plantes succulentes' could enrich the soil over which they grew.[2] Here again, examples came from England, in books such as *The Improver Culture*, advertised in the *Journal de Physique* (1777). But the confused knowledge of chemical conditions prevailing at the time, narrowly limited the scope of research in that sphere.

Finally a third means of enriching the soil and supplying food for plants was found in what the Abbé Toaldo called 'météores'[3]. This author remarked that 'le grand Newton pensait que les plantes absorbaient outre l'air et l'éther, les particules du feu et de la lumière. Monsieur Franklin et d'autres physiciens sont du même sentiment.' Tull and Duhamel had already noticed that repeated ploughings allowed water and various elements, which they deemed indispensable for the food of plants, and which were contained in the ambient air, to penetrate the soil. Home had stated that 'la terre vierge exposée au vent du Nord renferme une plus grande quantité de principes propres à donner l'essort à la végétation que tout autre terre'.[4] But it was only towards the end of the period that the experiments pursued by Priestley, Lavoisier and Ingen-Housz led this part of agricultural chemistry towards new discoveries. The evolution of this movement, however, was in a sense associated with the new agriculture, since application of meteorology to agriculture was inaugurated by Duhamel and led[5] to the rather remarkable study of Toaldo, in which this question is closely linked to the new agricultural system.

[1] *Traité*, vol. I, ch. IV, pp. 26-43.

[2] *Eléments*, vol. I, 'Des engrais que fournissent les végétaux', p. 187.

[3] 'Essai de météorologie appliqué à l'Agriculture' (Ouvrage qui a remporté le Prix de la Société Royale des Sciences en 1774 sur cette question: Quelle est l'influence des Météores sur la végétation? Et quelles conséquences pratiques peut-on tirer relativement à cet objet, des différentes observations météorologiques faites jusqu'ici? Par l'Abbé Toaldo, Professeur d'Astronomie, Géographie et Météorologie à l'Université de Padoue), *Journal de Physique*, 1777.

[4] Review in *Nouvelliste Oeconomique*, 1757, vol. XX, pp. 3-8.

[5] After such books as M. de la Folie's *Le Chimiste et l'Agronome*, and other meteorological studies by Abbé Dicquemare and du Luc.

A word should be added, however, on the future of the new science. In 1789 the discoveries of Lavoisier had only begun to be accepted by the scientific world.[1] And whereas his predecessors in spite of their works and in spite of many an ingenious theory can only be considered as experimentalists, Lavoisier can undoubtedly be called the father of the modern system of chemistry. Indeed, although his discoveries only reach agricultural questions by the side way of vegetable physiology, a remarkable group of scientists continued his work, and prepared the way for the great agricultural chemists of the nineteenth century, Liebig and Boussingault. Among those were Fourcroy and, above all, Chaptal. Fourcroy, like most of his contemporaries, was mostly interested in vegetable chemistry, but he widely assessed the importance of the new science when he wrote: 'Dans l'étude des propriétés des végétaux, les applications chimiques ou les données de la chimie sont seules capables de porter la lumière que les observations des naturalistes, malgré leur grand nombre et leur sagacité, n'ont pas pu y porter encore: elles seules peuvent éclairer la physiologie végétale et l'agriculture comme tout les découvertes modernes le prouvent aujourd'hui.'[2]

Speaking of the history of the chemical analysis of vegetables he recalls the unsuccessful attempts of the old chemists and shows how it is only through the discoveries of Black, Priestley, Ingen-Housz and Sennebier that vegetable compounds are constituted by these elements, carbon, hydrogen and oxygen.[3] He fully understood the consequences of the new chemical notions on the nature of air and on the composition of vegetable matter. 'Toutes ces merveilles de la culture qui varient et multiplient nos jouissances, s'expliquent ou se conçoivent par les notions chimiques qui ont

[1] See in *Annales de chimie*, 1789, a critical review of Kirwan's *Essay on phlogiston*, vol. VII.

[2] Fourcroy, *Système des connaissances chimiques et de leurs applications aux phénomènes de la nature et de l'art*, Paris, an. IX, vol. I, p. XXXV.

[3] *Ibid.* vol. VII, p. 71.

été développées dans les articles précédents'[1] and he shows how much the constitution of a *real* agricultural chemistry owes to Lavoisier: 'Wallerius avait déjà fondé quelques bases sur la chimie agricole; Bergman les avait poussées un peu plus loin en s'occupant des terres; Lavoisier en avait fait depuis ressortir toute l'importance.'[2]

But Chaptal, who under the Restoration was to apply himself with success to agricultural chemistry, can be considered as the first real agricultural chemist in France. In 1791 already, the first words of his Elements of Chemistry were significant enough: 'Agriculture is no doubt the basis of public welfare because it alone supplies all the wants which nature has connected with our existence. . . . The cultivation of the arts is therefore become almost as neccessary as that of the ground; and the true means of securing these two foundations of the reputation and prosperity of a state consists in encouraging the Science of Chemistry which discovers their principles'[3]: he adds further 'Agriculture is more intimately connected with chemistry than is usually supposed. It must be admitted that every man is capable of causing ground to bear corn; but what a considerable extent of knowledge is necessary to cause it to produce the greatest possible quantity!'[4] He was also among the first to emphasize strongly the necessity of providing nitrogen to cultivated plants.[5] But it was not until 1840 and the publication of Liebig's treatise of organic chemistry in French[6] that the relations between the chemical composition of plants and the food they required were to be fully understood.

Actually the connections between this movement in France and

---

[1] *Ibid.* vol. VII, p. 320.

[2] *Ibid.* vol. VII, p. 321.

[3] Chaptal, *Elements of Chemistry, translated from the French*, London, 1791, vol. I, pp. i and ii.

[4] *Ibid.* p. xvii.

[5] *Ibid.* vol. II, p. 31.

[6] Justus Liebig, *Traité de chimie organique*, Paris, 1840.

the movement which ran parallel in England at the same time[1] show, however, that in the case of vegetable chemistry France, in turn, was taking the lead.[2] But it can be said that during the pre-Lavoisier period, the role of British scientists in the orientation of researches into the field of agricultural chemistry must not be underestimated. Briefly, although a direct influence from England is noticeable only in the case of manures, and more especially, of mineral manures, it is none the less true to say that, on the inception of the new agricultural movement, agricultural chemistry as a whole began to develop. The major element in modern agriculture was thus inaugurated.

[1] Made famous by the names of William Henry (*An Epitome of Chemistry*, London, 1803), Kirwan (*op. cit.*), C. R. Aikin (*An account of the most recent discoveries... in chemistry*, London, 1814) and finally Sir Humphrey Davy (*Elements of Agricultural Chemistry*, London, 1827).

[2] This was not the case however for metallurgic chemistry. See Guyton de Morveau, *Eléments de chymie théorique et pratique*, Dijon, 1777, vol. 1, pp. 248 ff.

PART FIVE

SOME ASPECTS OF THE INTERNAL LIFE OF THE AGRONOMIC MOVEMENT

CHAPTER XII

HOW THE AGRONOMES OBTAINED THEIR INFORMATION ABOUT ENGLAND
1750-1789

AFTER an attempt to point out the main fields of the agronomic literature, in which the influence of English agriculture is particularly noticeable, it is now time to show how this knowledge was practically acquired, how the interest which had manifested itself even before 1750 was maintained and increased with the Revolution, in what manner the agronomes made contact with British specialists, and which were the channels they used—in one word, the framework on which the movement of agricultural research rested. Marc Bloch very rightly said that studying the history of a technique, means studying that of contacts between minds.[1]

The knowledge of the agronomes was not derived exclusively from books. Many actually knew the country of which they were speaking, or had at least come in contact with its inhabitants, in one way or another.

It seems that England was better known than some agronomes realized.[2] Towards the end of the *ancien régime*, travel to England became one of the rules for cultivated Society. Following the example given by the English themselves, the French more frequently undertook the crossing of the Channel. Fashion created closer links between the aristocratic circles of the two countries,

---

[1] M. Bloch, *Les caractères originaux*, p. 222.

[2] 'M. le duc de Liancourt a annoncé qu'on connaissait peu l'agriculture anglaise. . . .' Pigeonneau et de Foville, *op. cit.* p. 207.

so that a better understanding of both civilizations became possible. The propaganda spread by anglomane travellers in France had positive results, and account has been taken of these exchanges in the fields of literature, politics and society. Very little, however, has been said about its economic consequences, and practically nothing exists from the agricultural point of view. In the superficial lists drawn up by Mr Lockitt or Miss C. Maxwell[1] we can find one or two names, vaguely related to the economic activity of the period: Adam Smith for instance and the inevitable Arthur Young, who might almost seem at times to have been the only agricultural traveller of his day.

In fact, however, between 1750 and 1791, there is a continuous exchange of agronomes and agricultural scientists between England and France, which enables us, to some extent, to check up the conclusions of French agricultural literature, or at least to explain the diffusion and fashion of English methods in those times. Detailed accounts of these travels, Young's excepted, hardly exist as such. The agronomes were very seldom literary men or letter-writers, as was the case among members of the aristocracy, or of the philosophical and literary worlds; but it is

---

[1] Social and intellectual contacts only have so far been studied. Besides the works of Weulersse, Carcassonne and Bonno which analyse the English influence on economic and political ideas, the main studies are concerned with literary exchanges (J. Texte, *J.-J. Rousseau et le cosmopolitisme littéraire*, Paris, 1898. F. C. Green, *Minuet*, London, 1939, etc.). There is very little to be found in Corneille de Witt, *La Société anglaise et la société française au XVIIIe siècle* or de Rémusat, *L'Angleterre au XVIIIe siècle*, Paris, 1857. The old E. J. B. Rathery, *Des relations sociales et intellectuelles entre la France et l'Angleterre*, Paris, 1856, may, however, still prove useful. But neither of these books actually attempts to rate the movement of travellers, for instance, or to estimate to what extent personal contacts were established. There are only rather slight works on this subject: H. Lockitt, *The Relations of French and English Societies, 1783-93*, London, 1926. C. Maxwell, *The English Traveller in France, 1750-1815*, London, 1932. The unfinished work of Mathorez, *Les Etrangers en France*, Paris, 1919, 1921, does not unfortunately contain the part he intended to present, on the English in France. Details can be found in A. Babeau, *Les Voyageurs en France depuis la Renaissance jusqu'à la Révolution*, Paris, 1885.

possible, here and there, to discern a picturesque detail among all the specialized material they produced. From England, the agronomes brought back facts and technical observations which they included in their scientific writings. It is therefore less the actual description of the journeys which count than the fact that agricultural travels were undertaken between the two countries, which greatly promoted the agronomic movement in France.

The period immediately before 1750 may be considered as the prelude to the more intensive movement of travel which follows. It was not a question yet of agricultural tours. It was, as we have seen, a virtual discovery of England, of its various aspects, its literature, its government and its countryside. To the French in England, as well as to the English in France, agricultural observations were only a supplement to more pleasant and more popular descriptions. Meanwhile closer contacts were established between the scientific worlds of the two nations. The English seem to have been more active, at first, than the French. The traditional 'grand tour' bore fruit in some instances.[1] It has been said that Tull, for example, owed much to Languedoc for its confirmation of his new theory.[2] One of these 'tours' was the reason of Buffon's journey to England, and it is possible to assign to it an important influence on the future career of the naturalist.[3] But as it is only around 1750 that the first noticeable results of the improvements undertaken twenty years before appear in England, it would be artificial to find any agricultural motives for travels before this date. Nevertheless, some of them were indirectly connected with agriculture; as, for instance, those made by Du Fay and Duhamel to study timber.

More important for the future of French agriculture was the

---

[1] For instance, Turgot had the opportunity of meeting Adam Smith at d'Holbach's in 1766, when the English economist was travelling as tutor to the Duke of Buccleuch; see D. Dakin, *Turgot*, London, 1939, p. 17.

[2] Lord Ernle, *op. cit.* p. 171.

[3] Condorcet, *Eloge de Buffon*, Paris, 1847, vol. III, pp. 327, 329.

settlement in France of many Englishmen driven there because of their political convictions. Some of them belonged to the Jacobite colony established in France in the time of Stuart exile.[1] It would no doubt be easy to find among them gentlemen interested in agriculture and propagating in France the methods of their own country. Lord Ogilvy and Henry Patullo were the most important representatives of the Scotchmen who had a strong influence on the French agronomes. We can see in Michel Adamson's Scotch origin an explanation of his botanical fortune. After 1745 and the failure of the Stuart attempt, more refugees arrived in France.[2] Among them were technicians whose influence was to be very great, chiefly in industry,[3] but also in agriculture. From this time up to the end of our period, English colonies of various size may be found in certain regions of France, and they will not be without influence on the agricultural destinies of the surrounding country. An interesting instance is that of La Réveillière-Lépeaux, the future member of the Directoire, who was an enthusiastic botanist.[4] Well acquainted with the English language, he was also familiar with English thought[5] because he lived in Angers where there was a famous Riding School which attracted a great number of British gentlemen; and we know that some of the tutors of the young aristocrats were well worth

[1] See Marquise de Campana de Cavelli, *Les derniers Stuarts et la Cour de St. Germain en Laye*, Paris, 1871. Other details in B. Fay, *La Franc-Maçonnerie et la Révolution intellectuelle du XVIIIe siècle*, Paris, 1935.

Arthur Young mentions a M. de Galway of Irish descent, who bought Turbilly's estate, *Travels*, vol. I, p. 252.

[2] The English, Scottish and Irish Regiments in France were not without influence on the question. Among their officers were Ogilvy and de Mante. See *Histoire des troupes étrangères au service de France*, par E. Preffe, Paris, 1854, vol. I.

[3] Like Holker. See P. Boissonade, 'Trois Mémoires relatifs à l'amélioration des manufactures de France sous l'administration des Trudaine, 1754', *R.H.E.S.* 1914, vol. VII, pp. 56-86.

[4] *La Réveillière-Lépeaux, Citizen Director, 1753-1824*, by Georgia Robinson, Columbia University, Ph.D. thesis, New York, 1938, pp. 38, 39, 58, 59 *passim*.

[5] *Biographie*. Miss G. Robinson ignores the fact.

meeting.[1] But the most definite example is that of Boulonnais where improvements, begun around 1770, were an actual consequence of direct English influence. In a picture of this country, an agronome wrote in 1812:

> Il y a cinquante ans que l'état moral et physique du Boulonnais éprouva des changements importants, quoique insensibles dans les premières années, lorsque la grande route de Paris à Calais (qui passait autre fois par St Omer) le traversa dans toute sa longueur, et que d'autres routes furent ouvertes sur d'autres points. Des voyageurs nationaux et étrangers qui allaient en Angleterre ou qui en revenaient, résidaient momentanément dans ce pays ou se fixaient dans ces villes; l'on doit attribuer au séjour et à la fréquentation de ces voyageurs les lumières qui commencèrent à se répandre dans l'arrondissement; et Boulogne en particulier qui ne comptait auparavant que des propriétaires et des pécheurs vit sa population se composer de négociants, d'hommes instruits et de quelques capitalistes empressés de rejeter sur leur propriétés des essais heureux et des découvertes utiles.[2]

Without doubt the transformation of Boulonnais noted by M. Bloch[3] was stimulated by the English influence so clearly shown here. Perhaps the changing of the high road quoted is responsible for the marked difference still noticeable at the end of the eighteenth century between the grassy regions of Flanders and Boulonnais, and the backward state of Picardy.[4] Arthur Young and Dr Rigby also point out the existence of British colonies in Calais and Boulogne. These may have influenced in some ways the French

---

[1] Another famous *Académie* of riding existed in Caen. Normandy was much visited by English tourists (Evelyn, Peter Heyliss, Philip Thicknesse) and penetrated by the influence of England, mostly, it is true, in the literary field. See Paul Yvon, *Traits d'union anglo-normands*, London, 1919. But we find in 1775, an English Officer, De Mante (an agronome himself, author of a treatise on *L'Education des moutons de race anglaise*, Paris, 1778) who was given by the French Government a huge space of barren lands near Arques, on condition that he should enclose them and grow artificial grasses. In J. Sion, *Les Paysans de la Normandie orientale*, Paris, 1909, p. 212.

[2] *Mémoires d'Agriculture*, 1812, vol. xv, p. 354.

[3] *Loc. cit. A.H.E.S.* 1930, p. 328.

[4] Arthur Young, *Travels* (edition Sée), vol. ii, p. 549.

Society in these same towns. Calonne thinks that the longer leases in the Calais region, for instance, can be attributed to English influence.[1]

Such is one aspect of English example. Stronger and more extensive was that which the travelling agronomes brought back from England itself. It is after 1750 that journeys with a definite agricultural purpose begin. Those who so travelled were not brilliant personalities, eager to acquire literary glory through their knowledge of England. They were mostly scientists, or *propriétaires*, or administrative officials. Their impressions are not to be found in witty remarks or picturesque notes, but only as a part of the factual accounts of their personal undertakings. The most striking example of the documentary journeys is that of Roland de la Platière which was to have important consequences in the research of livestock. Roland states his eagerness for information and decides to go to England, 'pour y visiter les troupeaux de bêtes à laine, les terrains sur lesquels ils vivent et y étudier les pratiques relatives à l'éducation de ces animaux et à la perfection de leur laine'.[2] The time was most unsuitable, as war with England was at full blast, and it was impossible to cross the Channel by official means. And it is with genuine pride that Roland announces that he risked his safety for so important an undertaking: 'l'importance de celui-ci m'a fait fouler aux pieds les dangers auxquels de semblables démarches exposent, dans un temps surtout où la frayeur de se mettre en mer était égale de part et d'autre à cause des hostilités commencées, qui ne m'ont permis en outre que de mettre en usage que des moyens très périlleux.'[3] It was English methods of stock-breeding which for the most part attracted the French gentlemen stock-breeders (travel to England being a privilege of the rich, and of those enlightened

---

[1] Calonne, *op. cit.* p. 66.

[2] 'Mémoire sur l'éducation des troupeaux et la culture des laines', *Journal de Physique*, 1779, vol. XIV, p. 64. [3] *Ibid.* p. 63.

landlords who on returning to their lands were supposed to propagate the English agricultural gospel); these men could easily visit the famous estates of the country, and negotiate the purchase of some fine specimens of animals. One of these men was M. de Lormoy, a big grazier of Boulonnais, and therefore well acquainted with English methods. He tells how in 1760 he smuggled 'une assez grande quantité de bêtes à laine'.[1] So great is the importance attached by Lormoy to English methods, that he criticized Broussonnet, Secretary of the Royal Society of Agriculture, for ignoring the fact that sheep ought to feed mostly on turnips: 'Il ne l'ignorait sûrement pas s'il avait été en Angleterre ou s'il avait eu des notes fidèles à ce sujet.'[2] But he was misinformed. Indeed, Broussonnet had been to England in 1784, and there met Faujas de St Fond en route for Scotland.[3] The Duc de Liancourt had crossed the Channel in 1768 and spent his time visiting farms, and getting information about soil cultivation, cattle breeding and so on.[4] He found the country so interesting that he sent his sons later with Lazowski as their Tutor, thus imitating the custom of the English aristocracy.[5] One of the most interesting practical husbandmen of the time, Delporte, also sent his sons to England to be educated and to learn the management of an estate à l'anglaise.[6] Another agronome from Boulonnais, Dumont de Courset,[7] was visiting England in 1783. Back in France, he published a *Mémoire sur l'Agriculture du Boulonnais et des cantons maritimes voisins*, in which he lays down in the light of his English experience how the region ought to be improved. From Ile de France, the Marquis de Guerchy also went to

---

[1] De Lormoy, *Mémoire sur l'Agriculture*, Paris, 1789, pp. 46-7.

[2] De Lormoy, 'Instruction sur les turnips', *B.P.E.* 1786, vol. II, p. 124.

[3] Faujas de St Fond, *A Journey through England and Scotland* (edition with notes and a memoir of the author by Sir A. Geikie), Glasgow, 1907, vol. I, p. 16.

[4] *Vie du duc de la Rochefoucauld-Liancourt, par son fils*, Paris, 1827, p. 15.

[5] F. de la Rochefoucauld, *op. cit.* Introduction, p. 12.

[6] See *Biographie*; also Roland's report in *Journal de Physique, op. cit.* p. 100.

[7] *Biographie.*

England.[1] This indefatigable propagandist of the *éducation sauvage* and, in general, of all English agricultural methods, brought back valuable notes on his experiences which he used in his writings on sheep. Horse-breeding, so famous in England, also attracted observers and customers from France. Members of the Court aristocracy sent business men over the Channel to purchase at great expense riding-horses, race-horses and famous stallions.[2] Many of the greatest nobles did the journey themselves, like the Count of Lauraguais,[3] the Duc de Chartres[4] and the Duc de Lauzun.[5] Newmarket and, in general, all the famous stud-farms were frequented by a quantity of French gentlemen whose influence on horse-racing and horse-breeding in France was to be very important.[6] Some time later, Bakewell and Coke of Holkham also received many foreigners, among whom may be found people like François de la Rochefoucauld and Baert, who attended their famous agricultural sessions,[7] and marvelled at the results obtained on their estates.

Besides these leading names of the new French agriculture, anonymous signs of a practical knowledge of English methods are innumerable. Duhamel obtained first-hand information about Tull's methods. The articles which report agricultural customs of various English counties in French periodicals cannot be numbered. Relations between pure scientists of the two countries increased as the century advanced. Some of their travels helped a great deal in the introduction into France of English discoveries. La Condamine is the typical example of these members of the Academy of

---

[1] *Mémoire pour l'amélioration. . . ,* Paris, 1785, *loc. cit.*

[2] Robert Black, *Horse Racing in France,* London, 1886, pp. 10-11.

[3] See below, p. 209, n. 1.

[4] A. Britsch, *Jeunesse de Philippe-Egalité,* pp. 394, 395, 398. Duport de Cheverny went to England in 1750 to purchase carriage horses, in Carré, *La noblesse de France,* Paris, 1920, p. 206.

[5] A. L. de Gontaut, duc de Biron, *Mémoires (1747-83),* Paris, 1858, p. 132.

[6] R. Black, *Horse Racing in England,* London, 1893, p. 87; *Horse Racing in France,* London, 1886, pp. 8-9.   [7] Ernle, *op. cit.* p. 185.

Sciences who undertook the journey to England not, as a matter of fact, in order to study agriculture, but to 'voir des hommes'.[1] Bernard de Jussieu crossed the Channel twice and brought back one of the precious conquests of French arboriculture, 'le cèdre du Liban qui manquait au Jardin du Roy'.[2] Faujas de St Fond came to this country with the intention of studying the minerals of Great Britain. In London, he visited Sir Joseph Banks who gave him some precious seeds of chinese hemp, which he had just received. After he returned to France, Faujas distributed the seeds to several naturalists and agronomers and, later, any one could obtain them. The first beneficiaries of the new crop were 'MM. de Malesherbes, de Rosambo, Trudaine l'aîné, Boutin, Lavoisier, Hell, Varenne de Fenille, Buffon, Thouin, Moral'.[3]

Thus penetrated into France not only a direct knowledge of English agriculture but also material acquisitions from its results. Another and no less interesting aspect of this penetration is the deliberate action of the French Government to acquire information on this agriculture in order to improve that of France. The period 1750-1789 is that of reforming ministers, interested in new economic theories. Some of them, like Maurepas, Trudaine,[4] and Turgot,[5] particularly attracted by English methods, and following the general tendency, began to learn English.

---

[1] Condorcet, *Eloge de la Condamine*, p. 199.

[2] Condorcet, *Eloge de Jussieu*, pp. 267-8. Another important 'trait d'union' between the two countries was the seed-merchant Vilmorin whose influence was great in the question of trees for pleasure gardens. See 'Eloge de Vilmorin', *Mémoires d'Agriculture*, vol. x, p. 310.

[3] Faujas de St Fond, *Journey through England and Scotland*, vol. i, p. 16, note. Also *B.P.E.* 1786.

[4] Condorcet, *Eloge de Maurepas*: 'Il étudia la langue anglaise presque inconnue en France à l'époque où il aurait pu l'apprendre, mais devenue pendant son ministère, la langue étrangere la plus cultivée', vol. ii, p. 487. Also, *Eloge de Trudaine*, vol. ii, p. 236.

[5] D'Argenson also pointed out the influence of Dangeul's book (*op. cit.*) on Machault's economic policy. (*Journal and Memoirs*, Edition Rathery, vol. viii, pp. 299-300.) See M. Marion, *Machault d'Arnouville*, Paris, 1892.

This action of the French Government was twofold. First, they tried to obtain detailed information, statistics, descriptions and explanations of the state of agriculture in England. The memoirs thus produced helped in enlightening the Administrative Boards concerned with agriculture. Secondly, they officially encouraged the purchase (even if illegal) of English products in order to enrich and support French research on the subject.

Very often, certain pressing requirements made this information particularly urgent. The Government were led at times to consider some agrarian problem and looked for its solution in England. Such was the question of enclosures and abolition of common pasture. Such was also the question of common lands.[1] An enquiry was made to M. de Vatan, Intendant of Caen, who was requested to 's'informer comment les Anglais rendent leur terres incultes utiles à leurs bestiaux, cette nation retirant un grand avantage de ces terres'. The Intendant answered that nobody having been able to provide any information about it, he had charged a merchant to enquire on the spot, and would send the report to the Government.[2] Other unforeseen circumstances, like the fodder dearth of 1785, compelled the Government to immediate action. Precise and first-hand information had then to be obtained. M. Bloch has shown how such enquiries were addressed to the Minister in charge of agriculture, Bertin.[3] Other problems like those of woollen factories and wool weaving, in close connection with stock-breeding, also provoked governmental action. Such information as that provided by the French Embassy in London[4]

[1] H. Sée, 'La Mise en Valeur des terres incultes . . . à la fin de l'Ancien Régime', *R.H.E.S.* 1923, pp. 62 ff. *France Economique et Sociale*, op. cit. p. 60.

[2] H. Sée, *La Vie Economique* . . . , op. cit. p. 96, n. 1.

[3] He quotes (*A.H.E.S.* 1930, p. 355, n. 2) a 'Mémoire de Guerrier, associé de la Société d'Agriculture d'Alençon' on 'culture anglaise' sent to Bertin in 1766 and analysed in the Comité d'Agriculture. This Guerrier must have been the brother of M. de Lormoy.

[4] Enquiry sent to the Duc de Nivernais, French Ambassador in London (1763), *A.H.E.S.* 1930, p. 354.

or by English refugees, like Holker,[1] was to influence enormously the solution of the question. It would be a matter for research to attempt to weigh the actual amplitude of these enquiries, and to know the number and quality of the persons employed. Among them, many must have belonged to the agronomic world. We have already met Duhamel in official missions,[2] and the Abbé Morellet who travelled and made enquiries in England at the expense of the Caisse du Commerce.

At the beginning of the nineteenth century, De Pradt undervalued the action of the *ancien régime* government when he urged the Consuls (1802) to undertake and give the utmost publicity to 'tout ce qui se publie en France et à l'étranger sur la culture. Elle est l'objet continuel des recherches, des essais et des publications d'une multitude d'hommes éclairés en tout pays, et surtout en Angleterre.'[3] This is ignoring the actual efforts made by the old regime to obtain extensive documentation in England. There were for agriculture the same official enquiries as for industry, obtained from special envoys empowered, more or less, to spy out English processes.[4] For instance, under the influence of Daubenton and Tessier, there was a fruitful, although discreet, activity in the agricultural department of the Government. When it was a question of importing into France the best breeds of known sheep, Spanish merinos and English sheep, all sorts of means were employed. Diplomatic agreements with Spain were reached which

---

[1] See Boissonade, 'Mémoire tendant à multiplier et perfectionner les fabriques en France', *R.H.D.E.S.* 1914, p. 68. Also Ch. Schmidt, 'Les débuts de l'industrie cotonnière en France, 1760-1806', *R.H.D.E.S.* 1914, p. 26. On the role of English engineers in steel and coal problems see Levainville, *L'Industrie du fer en France*, Paris, 1922, pp. 48-58. Governmental enquiries were not limited to the field of industry. They were also extended in the same way to agriculture.

[2] Maurepas also sent, on Duhamel's recommendation, M. Olivier to England to study timber for naval purposes. Condorcet, *Eloge de Maurepas*, vol. II, p. 476.　　　　　　　　　　　　　　　[3] De Pradt, *op. cit.* p. 27.

[4] *R.H.D.E.S.* 1912, pp. 29 ff. Arthur Young gives interesting details on Englishmen employed in French manufactures, *Travels*, vol. III, pp. 955-76, *passim*.

authorized (rather parsimoniously indeed) the entrance into France of a merino flock, while others were officially smuggled across the border.[1] The problem was more difficult with England. Very strict laws prohibited the exportation of sheep,[2] and in order to circumvent them, the French Government was obliged to act as a private individual and pay for their smuggling by sea. In 1763 the smuggling of rams for reproduction was decided upon by Bertin. Helped by the zeal of Intendant Maynon d'Invau and the Prince de Croy, he succeeded in obtaining three Lincolnshire rams and six ewes. The animals were transported by the Post-master Caffieri, who landed in June between Sangatte and Calais.[3] In Bacalan's observations in 1765 we read that Boulogne was the headquarters of smugglers between France and England.[4] The sheep which had been so obtained were sent to the Royal Stables of Rambouillet, the director of which was Tessier, or given to Daubenton, who had established model stables at Montbard in Burgundy. Official propaganda was launched about the new improved breeds and the Contrôle Général sent letters to the Abbots of agricultural Orders to recommend the use of English sheep.[5] In 1786 when Bakewell's results began to be better known and praised in the official institutions,[6] M. de Minut was able to state at the semi-official Comité d'Agriculture that the

---

[1] See Tessier, 'Bergeries Nationales', *Annales de l'Agriculture Française*, 1831.

[2] '. . . M. de Lormoy avait trop ébruité en Angleterre le projet d'en tirer des bestiaux, qu'il venait d'être rendu dans ce royaume une loi qui défend d'exporter des bestiaux vivants', in Pigeonneau et de Foville, *op. cit.* p. 359. Export of wool was also severely punished. E. Lipson, *The History of the Woollen and Worsted Industries*, London, 1921, pp. 87 ff.

[3] 'Mémoire concernant l'achat de brebis et de béliers en Angleterre et leur transport en France', *Archives de la Somme*, chs. 26, 8, 12, see Calonne, *op. cit.*

[4] *R.H.D.E.S.* 1908, p. 394.

[5] See Levasseur, 'De Progrès de l'Agriculture', *Académie des Sciences Morales et Politiques*, 1898.

[6] Flandrin, professor at the Veterinary School of Alfort, had much appreciated the records of Bakewell after his journey to England. See Lavergne, *La Société d'Agriculture de Paris*, Paris, 1870.

Government was making considerable sacrifices in order to import sheep from England.[1] It is therefore certain that practical, though intermittent, action had been undertaken by the Administration in order to get material proofs of English agricultural achievements.[2]

Smuggling of sheep was not the only object of governmental attention. In 1763, when the potato began to be better known, M. de Boyne, Minister of the Navy, charged M. Chanlaire, a navy official in Boulogne, to receive from England a cargo of potatoes. But these arrived in France in a sprouting condition, and could not be sent to the agronome Minister. They were distributed on the spot and were henceforward cultivated with increasing success.[3] Turnip seeds were also officially purchased in England in 1786.[4]

A word should also be added about the influence exerted on agriculture by such important bodies as Provincial States or Provincial Assemblies.[5] There, big *propriétaires* were often influential and some were actually aware of the new agricultural trends. This was reflected in the proceedings of these bodies or their economic recommendations.[6]

The various societies which in the second half of the eighteenth

---

[1] In 1788 the Provincial Assembly of Normandy tried to smuggle Lincolnshire sheep. But the English Navy led to the failure of the scheme; see Sion, *Les Paysans de la Normandie orientale*, p. 243.

[2] We only speak here of the attempts made to improve French farming. In other days cattle had been imported from England and Ireland in order to prevent a possible meat shortage and lower the prices (see Babeau, *La Province sous l'Ancien Régime*, Paris, 1894, p. 257).

[3] *Mémoires d'Agriculture*, vol. xv, 1812, pp. 423-4. Mustel wrote in 1770 that potatoes were so scarcely known in Paris that he was obliged to order them in England in order to introduce them in Normandy. Sion, *op. cit.* p. 253, n. 1.

[4] Pigeonneau et de Foville, *op. cit.* p. 5, n. 1.

[5] For further details see M. Bloch, *loc. cit. A.H.E.S.* 1930, pp. 347, 354; P. Renouvin, *Les Assemblées Provinciales de 1787*, Paris, 1921, pp. 215-23. See also L. de Lavergne, *Les Assemblées Provinciales sous Louis XVI*, Paris, M. Levy, 1864, which gives interesting technical details.

[6] Grimaux, *Lavoisier*, Paris, 1888, pp. 167 ff.

century gathered together individuals interested in farming problems, represent another means of information on the new agriculture. In a country which ignored scientific agriculture, this sort
of semi-official clubs may be said to have been both extremely
active and useful. Although their immediate results were rather
meagre, they held an important place in the history of French farming. It was within these Academies and Societies that discussions
about the new agriculture took place among specialists. They
were sometimes put in charge of official enquiries, they sanctioned
by their authority what was worth knowing, in short, they performed work which has not, so far, been justly recognized.[1] Had
they only aroused in the nation an interest of some kind or other
in agricultural questions, even an artificial one, they would have
been useful. Actually, as has been often pointed out, their
influence was little felt in the country as a whole. But in this
study they have been considered as strongholds of the new ideas,
and their activity in the theoretical field alone led to research
important enough for them to appear particularly significant to
our subject.

The oldest and most famous of these societies is the Academy
of Sciences.[2] In reality, its role in this matter was not deliberate,
and never even clearly realized. But indirectly it played a definite
part as a support to English agricultural ideas. The Academy was
the official connecting link between the scientific worlds of France
and Great Britain.[3] Its relations with the Royal Society of
London were very close. Geometricians and astronomers who,
since Newton, considered England as their second intellectual

[1] Besides the elder Lavergne's work (*La Société d'Agriculture de Paris*, Paris,
1870) numerous details will be found in L. Passy, *Histoire de la Société Nationale
d'Agriculture*, Paris, 1914; Labiche, *Les Sociétés d'Agriculture au XVIIIe siècle*,
1908. Provincial Societies have generally been studied in local monographs or
periodicals.
[2] Maury, *L'Ancienne Académie des Sciences*, Paris, 1864.
[3] See list of foreign members of both Societies (Royal Society of London and
Académie des Sciences) in Thompson, *op. cit.* and in Rozier, *Table Raisonnée*.

country, were not the only ones to consider English science favourably. In 1745 Réaumur wrote to Mr Folkes: 'I heartily wish there was in the world as strong a moral attractive power as there is a natural one, that might dispose our two nations particularly to seek to unite by mutual acts of friendship and good will.'[1] In the second part of the century this 'attractive power' was to become a reality. A man like Lavoisier, for instance, did much to keep excellent relations with his English colleagues, Banks, Blagden, Priestley and Black. He recommended to them the Frenchmen who visited England or Scotland, and in the same way received at the Arsenal the English who were sent to him.[2] This regard for English science is not perhaps unconnected with the predilection for the development of agricultural science. In the second half of the century, interest for agronomy was in a sense crystallized within the Academy.[3] But until a seat was created for an agronome[4] it was botanists and physicians like Duhamel, Valmont de Bomare, Gilbert, Tessier, Broussonnet, who repre- sented agriculture in the Academy. At the end of the century this latter appeared as a supreme tribunal for all questions related to Science and, in fact, it supervised the work of an agricultural board like the Comité d'Agriculture.[5] Some of its members occupied agronomic key posts, such as the 'Jardin du Roi', or the 'Bergeries Royales'; some were sent on official missions to study acute agricultural problems.[6] In a word, the role of the

[1] Manuscript copy of Réamur's letter in the *Archives of the Royal Society.*

[2] In E. Grimaux, *op. cit.* p. 56.

[3] In 1753 the first paragraph of the *Règlement* states that agricultural observa- tions will be considered as 'preuves de capacité auxquelles le titre de correspondant doit être attaché'. See Weulersse, 'Le Mouvement Préphysiocratique', *R.H.E.S.* 1931, p. 249. Duhamel and Daubenton were the two Academicians who published a number of agricultural articles in the proceedings of the Academy. See, for instance, Duhamel, 'Observations Botanico-Météorologiques', in *Nouvelle Table des articles . . . de l'Académie Royale des Sciences . . . par l'Abbé Rozier*, Paris, 1775, vol. III, pp. 94-7.

[4] André Thouin, elected in 1786.   [5] Pigeonneau et de Foville, *op. cit.* p. 6.

[6] Duhamel and Tillet were sent to study a grain pest in Angoumois in 1760.

Academy in the agricultural revival, although undefined, was extremely important.[1]

The same part was played in the Provinces by local Academies, some of which had been expressly founded upon an English model.[2] In those also, agriculture belonged to the class of Sciences. They often doubled the work of agricultural Societies by offering prizes on agricultural research, and some of the enquiries thus raised remained famous or originated valuable works.

The new direction given to the work of the Academies has its origin in the imitation of what the English scientific societies had been doing since the beginning of the eighteenth century. Imitation of England was even more accentuated in the establishment of Agricultural Societies. Dangeul seems to have been the first to have suggested the creation of 'Sociétés de Cultivateurs', which, as in England, would transform agricultural methods.[3] The suggestion was rapidly welcomed in agronomic writings[4] and in 1757 the first of these societies was founded in Rennes.[5] 'C'est par une Société pareille que l'Irlande qui était une des plus pauvres contrées du monde est devenue très florissante. . . . Nous osons donc, Messieurs, vous indiquer un moyen qui a déjà réussi ailleurs

[1] A more restricted and specialized action, but still a very important one, was also undertaken, in connection with the agricultural revival, by the Société Royale de Médecine, which provided the necessary scientists for the study of epizootics, like Vicq d'Azyr and Tessier (this latter was sent on an official mission to Sologne to make a report on the cattle plague). See 'Observations sur plusieurs maladies des Bestiaux. . . . Par M. l'Abbé Tessier', in *Journal de Physique*, 1780.

[2] The Toulouse Academy of Sciences had been created in 1729, 'à l'imitation de la Société Royale de Londres'. *Histoire et Mémoires de l'Académie Royale des Sciences, Inscriptions et Belles Lettres de Toulouse*, Toulouse, 1782, vol. I, p. 2.

[3] 'Remarques sur les avantages et les désavantages de la France et de la Gde Bretagne. . . . Amsterdam, 1754.' (*Journal des Savants*, June 1754, pp. 175-6.)

[4] See Goudar, *Les intérêts de la France mal entendus dans les branches de l'Agriculture . . . par un citoyen*, Amsterdam, 1756. Also in La Salle de l'Etang, *Prairies artificielles ou Lettres à M. de . . .* , Paris, 1756. Mirabeau, *L'Ami des Hommes*, vol. v, p. 20.

[5] On the Société de Bretagne see H. Sée, *Les classes rurales en Bretagne. . . .* Paris, 1906, pp. 422, 423.

et dont le succès n'est pas douteux chez vous.'[1] This society, eager to compete with its Irish model, in 1759 entrusted M. Thébault with the translation of the Essays of the Society of Dublin. In 1757 it organized intensive propaganda for the new husbandry, recommended Bradley's treatise and purchased seeds in England.[2]

From 1757 to 1789 these Agricultural Societies were founded everywhere.[3] But their activity remained on the whole much too theoretical and speculative.[4] Mirabeau criticized them in his *Ami des Hommes* and urged them to follow the example of the Scottish Societies; they should give a prize only 'à celui qui fait rendre le plus de blé à son champ, qui tire le plus de croit de son troupeau'.[5] Modern authors point out that the influence of the Agricultural Societies declined after 1770.[6] There seem to be two reasons for that. To mention the practical one first: the Government did not continue to support them after Bertin, the Minister responsible for their creation, lost his influence.[7] At the same time, these Societies,

[1] 'Etablissement d'une Société d'Agriculture...', *Nouvelliste oeconomique*, 1757, vol. xx, p. 10. The example of the Society of Edinburgh was also advocated at this time. Later, the favour is accorded to the Society of Bath and the Society for the Encouragement of Arts in London.

[2] *Corps d'Observations*, 1757-8, pp. 101, 104, 137 etc.

[3] H. Sée, *La Vie Economique* . . . , *op. cit.* ch. I, 'Les Sociétés d'Agriculture, leur Rôle à la fin de l'Ancien Regime', pp. 5-24.

[4] Calonne, *La vie agricole sous l'Ancien Régime*, p. 42, thinks the main reason for the failure of the Societies was that their members were incompetent. It would seem more accurate to say they were too competent as they gave their agricultural research a very specialized and scientific turn, which the practical farmers could not possibly follow. See the arguments of H. Sée in *La Vie Economique* . . . , p. 17. An excellent account of the various activities of these Societies will be found in D. Dakin, *Turgot and the Ancien Régime in France*, London, 1939, ch. VI, which studies the Society of Limoges, pp. 79-91.

[5] The idea of giving prizes to farmers owed much to the English influence (agricultural competitions, later the 'Comices Agricoles'). Fresnais de Beaumont, an admirer of England, wished that farmers should be raised to the nobility, *La Noblesse cultivatrice*, Paris, 1778.

[6] H. Sée, *La vie économique* . . . , *op. cit.* pp. 5 ff.

[7] 28 May 1780. *Journal de l'Abbé de Véri* (edited by Baron Jehan de Witte), Paris, n.d. vol. II, p. 317.

instead of applying themselves to practical studies, emphasized the theoretical aspect of the new agriculture. Instead of promoting simple experiments, they raised very complicated problems, interesting ones no doubt, but problems which were bound to provoke the suspicions of the peasants. In a way, they did not know how to find a balance between local customs and the new tendencies. It is obvious that it is the scientific discussion about the latter which mainly interested them. They were then considered more as agricultural museums than as active bodies. The Intendant of Auch, although a member of the Agricultural Society, wrote these disillusioned lines: 'Il est douteux que cet Art [Agriculture] lui doive le moindre progrès dans cette province; peut être n'est-il pas plus redevable dans d'autres aux mêmes sociétés qui s'y trouvent établies?. . . Cependant il convient de maintenir ces associations; elles sont du moins un témoignage en l'honneur du premier et du plus utiles des arts.'[1] The creation of Agricultural Societies, then, seemed to have been a failure. The time had not yet come when the Provinces could find the spirit and the leaders necessary for an agricultural revival. At least the Societies had enabled certain amateurs to maintain a measure of activity through experiments and research. In 1785, however, the Paris Society was revived.[2] Thirty-five years had elapsed since the new technique of husbandry had been introduced into the country, since the first rather disordered enthusiasm. The new agriculture was being codified, and after a period of excitement about the fresh discoveries, the theory of a modernized French agriculture had been elaborated. The part played by this Society in the propagation of the doctrines of the final period was most important.[3] The proceedings which it published for four years contain some of the

---

[1] Quoted in Babeau's, *La Province sous l'Ancien Régime*, vol. II, p. 236.

[2] See Louis Passy, *Histoire de la Société nationale d'Agriculture*, Paris, 1912.

[3] Special mention must be made of one of its most active members, the Marquis de Guerchy, a great follower of English methods and a distinguished agronomic writer.

best examples of agricultural dissertations of the century.[1] In those, precise allusions to the state of things in England, genuine English principles, have replaced the somewhat romantic ideas of early times. Arthur Young was one of its guests and his severe criticisms completely underestimated its importance.[2] Together with the Royal Society of Agriculture, we must mention the Comité d'Administration de l'Agriculture, the history of which has been written. Even more than in the Royal Society, the Committee was a stronghold of English inspired tendencies represented by the Duc de Liancourt, Lazowski and Lavoisier.

The development of the agricultural movement in the second half of the eighteenth century is therefore explained when we consider how large was the stock of information available and how it was extended throughout France. It is therefore certain that as well as books, travels and societies, inspired by the love of agriculture, had a greater effect than they have been said to have had up to now. They created, in a way, the appropriate climate in which the new theories imported from England developed. They ensured that the *culture anglaise*, or the *nouvelle agriculture*, or the *culture de M. Tull*, should become the subject of research, of discussions, and of application. They, as well as the economist propaganda, caused the necessity of improving French farming by following English methods to be realized. They prepared the way to the large-scale changes of the following century. One last question remains to be asked. How did the timing of English propaganda vary between 1750 and 1789?

First of all, it was constant. Between Duhamel's book and Lazowski's memorandum, the interest of the agronomes was

---

[1] See praises addressed to it in the *Dictionnaire de l'Institut*, vol. 1, p. 19. Against it and after a comparison with the Society for the Encouragement of Arts, Commerce and Manufactures, see Chantreau, *Voyage dans les trois Royaumes d'Angleterre . . . fait en 1788 et 1789*, Paris, 1792, pp. 178, 179.

[2] He was elected a corresponding member on the 28th December 1785. See *Travels*, vol. 1, pp. 277, 278, 279. Criticism of his opinion by H. Sée, p. 279, n. 1.

constantly directed towards the study of agricultural modifications happening in England at the same time. At the moment
that France understood that grave changes were taking place, it
was still English agriculture which indicated the future way. De
Fresne's book,[1] after a comparative study of agricultural problems
in both countries,[2] showed that those of France were clearly
understood and interpreted. Forty years after the revelation of
English achievements, the same questions were still anxiously
asked:

Pourquoi par l'agriculture française, la terre est-elle moins fertile
que par l'agriculture anglaise? Pourquoi l'agriculture anglaise est-elle
dans toutes ses parties plus productive que la française? D'où viennent
ces désavantages en France? Est-il possible d'y remédier, quels en sont
les moyens? Suffirait-il pour cela d'imiter exactement les Anglais? et
le pourrait-on, le devrait-on?[3]

The agronomes were ambitious and their programme was far too
wide for wholesale application, even over a period of forty years.
The transformation of an agriculture like that of France into one
like that of England was a revolution in itself. But England was
there to show them that it was possible. Their fight for a better
agriculture had much of what Marc Bloch calls a 'couleur intellectuelle'.[4] This gap between their conceptions and the material
state of things led them not only constantly to complain, but also
constantly to seek to improve their researches and experiments,
while looking to England for new elements capable of giving their
theory the maximum efficiency.

The filiation of the agronomic school may be established as
follows. The typical representative of the first generation would
be Duhamel, who ended his career as a theorist of English

---

[1] *Traité d'Agriculture considérée tant en elle-même que dans ses rapports d'économie
politique, avec les preuves tirées de la comparaison de l'Agriculture, du Commerce
et de la Navigation de la France et de l'Angleterre*, Paris, 1788.
[2] See below, Appendix II.   [3] *B.P.E.* 1790, vol. I, preface, p. 5.
[4] *Les Caractères originaux*, p. 221.

husbandry. The second generation would be represented by Daubenton, or the Duc de Liancourt, who applied and diffused English methods with real success. The third one begins when our period ends. It is that of De Pradt, Thouin, François de Neufchâteau, Mathieu de Dombasle, who continued the tradition inaugurated half a century before. Finally, can it not be said that one of the most celebrated books on French agriculture of the last century, Léonce de Lavergne's *Rural Economy of France*, is in a way the successor of the agronomic books of the eighteenth century, which were also full of admiration of England?

CHAPTER XIII

PRACTICAL UNDERTAKINGS OF THE AGRONOMES

THERE is little more to be said on the deplorable condition of large estates. The offensive against absenteeism, waste lands on feudal domains, carelessness of the great landlords, was launched by the physiocrats and supported by agronomes like Patullo, who wished for a better cultivation of these neglected lands.[1] The example of England, where popular figures like Townshend, and his followers, had undertaken to obtain full value from their lands, was therefore proposed to the French aristocracy. Besides, it was the moment when political theoricians thought of giving the aristocracy a leading role in French economy. The second half of the eighteenth century is a time of a revival of the nobility, not in the sense of a selfish or blind nobiliary reaction, but a revival caused by the penetration into the nobility of liberal ideas, modelled on the English example. Lacretelle shows at the end of the *ancien régime* the ideal of the young aristocrats, as he must have witnessed it.[2] A new role therefore seemed to devolve on the aristocracy.[3] This did not include blood nobility only.[4] According to the new economic and political theories, only those who possessed land and money were the important element and the source

[1] See the famous comment of Arthur Young, *Travels*, vol. I, pp. 158, 159.

[2] Lacretelle, *Histoire de France*, Paris, 1809, vol. VI, pp. 22, 25, 32, 37.

[3] On this tendency of the young members of the aristocracy see Comte de Ségur, *Mémoires*, Paris, Londres, 1825, vol. I, pp. 130 ff.

[4] That is to say the upper classes. Indeed a *de* does not prove any nobility. It is, however, a sign of a tendency towards consideration. All our examples may not be taken from the category of the *gentilshommes*, but they belong in any case, to the classes in more comfortable circumstances.

of the country's welfare. Agriculture, therefore, could be restored only through *propriété foncière*, or *personnes opulentes*. The idea that land required huge investments of capital thus tended to the view that only through the rich *propriétaire* could the new methods of agriculture be established.[1] What the agronomes desired, was the creation of an aristocracy of money, of an agricultural capitalism, which would not only help the peasants by providing the necessary advances, but also take up model farming and act as an example.[2]

This was, in fact, a programme wholly English in its way. De Pradt, summing up this tendency, wished for 'l'enseignement de méthodes savantes et recherchées, qui exigent à la fois des études et des frais, des avances considérables en connaissance et en argent telles à peu près que les méthodes employées par les agriculteurs anglais'.[3] His programme, entirely aristocratic in inspiration, required that 'les méthodes savantes, imitées des étrangers, appartiendront aux hommes éclairés et riches, dont l'exemple formera au milieu du peuple, et sous ses yeux, un cours d'instruction pratique'.[4]

To an aristocratic society already imbued with English ideas, the programme was tempting. But we have seen what obstacles were opposed to it, and that the nobility followed the old traditional course of immediate profit.

Another tendency challenged the moneyed classes to invest capital into the land. Abbé Coyer's famous theory of the 'Noblesse Commerçante'[5] was less of a failure than at first appeared. Revived activity among the nobility is particularly noticeable in business,

---

[1] On a theory of social organization based on ownership of large estates and the criticisms of Roederer in his *Cours d'Organisation Sociale*, 1793. See E. Allix, 'Propriété foncière et fortune mobilière sous la Révolution', *R.H.E.S.* 1913, p. 305.

[2] This tendency is clearly attributed to the agronomes by Sion, *op. cit.* p. 211.

[3] De Pradt, *op. cit.* vol. I, Avant-propos, p. xl.

[4] *Ibid.* pp. xlii-xliii.

[5] E. Depitre, 'Le Système et la querelle de la Noblesse Commerçante, 1756-9', *R.H.E.S.* 1913, pp. 137 ff.

industry and financial speculation.[1] The tendency was to imitate the *bourgeoisie* in investing capital in factories, mines, colonies, etc., rather than the English aristocracy in developing agriculture, which was slow to yield returns. Fresnais de Beaumont's 'Noblesse Cultivatrice'[2] was less successful than the 'Noblesse Commerçante'. Besides, it was easier for the nobles, who possessed capital, to found an industrial company which could be directed from Paris, than to supervise their estates, which would require their constant presence in the provinces, away from Paris and the Court.

Impossible though it may be, however, to speak of a return of the capital to the land,[3] there are yet certain facts which should be fully mentioned, as being significant in this respect. There were enough individual cases of application of the agronomic teaching to make it permissible to speak of an agronomic movement among the nobility.[4] Always bearing in mind that enthusiasm may lead to a certain amount of exaggeration, we may yet accept the statement that the eighteenth century 'est devenu vraiment celui de la campagne, il a vu tous les hommes et surtout les citadins, s'y porter en foule, et le goût des maison de campagne s'est généralisé tout à coup, comme celui des jardins anglais l'a fait depuis'.[5] The rush to the country referred to above was a consequence of the interest awakened by the new agricultural theory and of the

[1] H. Carré, *op. cit.* pp. 250-63. Excellent views on the new economic trends within the aristocracy are found in M. Rouff, *Les Mines de Charbon en France, 1774-91*, Paris, 1922, Troisième partie, ch. I, pp. 174-208.

[2] He proposes the example of England 'qui voit dans les Fermes la moitié de sa noblesse.'

[3] Many authors point out that the *bourgeoisie* tended to invest its capital in lands (see Lizerand, *op. cit.* p. 153, and M. Marion, 'Les classes rurales en Bordelais au XVIIIe siècle', *Revue des Etudes Historiques*, vol. IV, 1902, pp. 108, 109 *passim*). I have not been able to find any evidence that this movement was determined by interest in agriculture. See Loutchisky, *op. cit.* pp. 150-1, Weulersse, *op. cit.* vol. II, p. 169.

[4] The facts which follow should be weighed against certain theories considering (even nowadays) the agricultural movement among the aristocracy either as a *bergerie* picture, often in the worst taste, or as dictated only by motives of greed.

[5] De Pradt, *op. cit.* p. 19.

possibilities it offered of increasing one's income.[1] 'Les cris qui se sont fait entendre à ce sujet ont été un éveil, une alerte pour les propriétaires et les possesseurs des terres. Chacun cherche aujourd'hui à améliorer ses possessions.'[2] This is no revolution, of course, but it is at least a movement towards improvement. In this more restricted sense, even the comparatively poor provincial nobility could do something. The movement was so clear that it was felt even abroad. The *Letters concerning the Present State of the French Nation,* state: 'There has lately existed and at present exists a spirit of improvement in matters of agriculture, amongst individuals in France which bears a countenance of great utility. ... A distinction should however be made between those worthy patriots, who labour to perfect the common practice and those who are chiefly busied in introducing new ones.'[3] This movement was noticeable in the highest sphere of the aristocracy also, where it was directly inspired by the English example. Some speak of a 'coterie agricole, commerciale et manufacturière qui allait chercher en Angleterre ses aspirations, ses goûts et ses idées politiques'.[4] Dupont de Nemours who was well informed on the subject, writes: 'Déjà beaucoup de seigneurs et des plus grands à tous égards, commencent à se livrer avec succès à l'agriculture qu'ils encouragent par leurs exemples, par le séjour plus long qu'ils font dans leurs terres, par des dépenses en améliorations qui d'abord ont paru être sur leurs revenus et qui n'ont pas tardé à accroître ces revenus mêmes.' Dupont was certainly alluding here to the La

---

[1] On the interest aroused in the young nobility by the physiocratic lectures, see Weulersse, *op. cit.* vol. I, pp. 134-5. Dedications of agricultural books also give interesting information. The most significant is Planazu's *Oeuvres d'Agriculture,* whose patrons are the Duchess of La Trémouille, 'Protectrice des Arts et de l'Agriculture', Viscountess de Pons, Prince d'Havré and de Croy, Prince Archbishop of Cambrai, Count of St Aldegonde, etc.

[2] *Jour. Oecon.,* 1763.

[3] Young, *Letters concerning the present state of the French nation,* London, 1769. Letter 11, pp. 41 and 43.

[4] Article: La Rochefoucauld-Liancourt, *Biographie.*

Rochefoucauld family, every member of which was at the time a strong supporter of the economic and agricultural movements. These gentlemen and the Duc de Béthune-Charost have since then been granted a certain documentary importance by modern historians. But they are far from being the only representatives of the 'grands seigneurs agronomes'.[1]

Between 1750 and 1761, during the period of testing the new system of agriculture, and before the temporary recultivation on a large scale of waste and barren lands, a great number of 'propriétaires' applied themselves to the improvement of their estates. Among them were Duhamel's correspondents. But the experimental movement seems to have been larger than this circle. The Maréchal de Noailles, who was responsible for the translation of Tull, made of his gardens a centre of experiments under the supervision of his gardener, Lebreton.[2] The Duke-King Stanislas undertook practical experiments on cultivation, at his country house of La Malgrange near Nancy, and grew artificial grasses and ray-grass. Lorraine was a province strongly affected by the new influence. In certain districts of it the enclosure movement was adopted. Its aristocracy was sometimes enlightened and hastened to create artificial pastures and perform experiments on their estates[3].

[1] Lavisse, *Histoire de France*, Paris, 1910, vol. IX, p. 217.

[2] In Cretté de Palluel, *Traité sur les Prairies artificielles*, Paris, 1801. See also Jussieu, 'Historique du Jardin du Roi', *Annales du Muséum*, 1805, vol. VI, p. 16, n. 1. L. Passy, *op. cit.* p. 226.

[3] Weulersse (*op. cit.*) quotes, 'En Lorraine, les seigneurs de la première distinction, les personnes les plus éclairées, s'empressent à l'envi de créer des prairies artificielles et de pratiquer des expériences sur leurs domaines' (vol. II, p. 158).
Contra: R. Parisot, *Histoire de Lorraine*, Paris, 1922, vol. II, 'La plupart des propriétaires se désintéressent des questions agricoles', p. 211. However, M. Bloch (*R.H.E.S. loc. cit.* pp. 348-540) speaks of 'influences novatrices' and H. Sée (*France Economique*, p. 43) shows how the enclosure movement was put into practice by great landowners. Besides such names as La Galaizière, Dom Miroudot (*op. cit.*), the Abbé Duquesnoy and de Commerell (in Parisot, *op. cit.*), Baron de Tschudi, translator of Miller (in his *Traité des arbres résineux et conifères*, 1768), and Lazowski, proves that the province never lacked experimental agriculturists who were to have a great successor, Mathieu de Dombasle.

Everywhere in the kingdom the influence of the new husbandry can be traced. The study of the feudal roll of Hauterive shows at this time that artificial grasses were cultivated and sometimes enclosed.[1] The account-book of a member of the Parliament of Rennes shows that in 1759 he undertook agricultural experiments and bought English clover.[2] In 1765 the *Foreign Essays* give the names of gentlemen who experimented on ridge-cultivation: MM. de Pontual, La Chalotais, de Laurencin, de Montluc. Patullo quotes 'exemples du produit des herbages artificiels' in the estates of 'M. Girardoz de Malassise, seigneur de la terre de Nandis près Melun, M. Le Clerc de la Varenne Ste. Hilaire, M. Quesnay le fils en Nivernois'. In 1768 the Abbé Baudeau praises the Prince of Rohan-Rochefort, 'qui ne néglige dans ses terres aucun moyen de bien faire'. Experiments on grain preservation were made by this same gentleman together with Berthier de Sauvigny, the Marquis de Puységur and Baron d'Espagnac de Puismarets.

The new husbandry also penetrated the ecclesiastical lands.[3] Sometimes intermediate agents for the new crop were heads of agricultural Orders, like Dom Miroudot in Lorraine, a specialist on English questions, and Dom Edouard Provenchères, Procurateur of the Chartreuse du Liget, near Loches.

Together with these examples of application of the new cultivation of lands, we find other practical husbandmen engaged in the improvement of sheep-breeding. It was there that the Abbé Carlier began his career as a sheep specialist. Experiments were made by the Marquis de Puismarais and the 'éducation sauvage' was tested at Chambord, estate of the Maréchal de Saxe.

Model estates began to be set up. The most famous one was that of the Marquis de Turbilly in Anjou, on which work had started

---

[1] *A.H.E.S.* 1938, pp. 302 ff.

[2] H. Sée, 'Un type de document; Le Livre de raison d'un Parlementaire breton au XVIIIe siècle', *A.H.E.S.* 1931, p. 232.

[3] See the action of Abbé Lefevre in Pigeonneau et de Foville, *op. cit.* Introduction, p. xviii.

in 1737. The agricultural prosperity of this district was partly due to the imitation of English methods.[1] Turbilly borrowed from the Agricultural Society of Dublin the idea of conferring on his peasants a sort of *Mérite Agricole* consisting of a sum of money and a medal—an idea which had also been advocated by Dupuy-Demportes. In Gatinais, at Denainvilliers, the brothers Duhamel had created a prosperous estate where land cultivation, nursery gardening and botanical gardening had attained a high level of perfection.[2]

Not very far from Denainvilliers, at Montbard in Burgundy, Buffon had assigned himself the part of a great landowner, and was experimenting on timber.[3] Thus an agricultural awakening seemed to be taking place among the *propriétaires*.

Between 1760 and 1770, the edicts on *défrichements*, partition of common lands, enclosures, actually determined another impulse of agricultural activity. *Défrichements, triage* of commons, enclosures, all were directed towards an enlargement of the estates and remained the privilege of the aristocracy, since they alone had enough money or land worth the undertaking. This revival had therefore been based on the *propriétaire* class. It sometimes adopted an entirely capitalistic outlook.[4] Companies were founded by capitalists who had been granted an allotment by the Government, and undertook the *défrichements*. The Company Vallet de Salignac and Chaulce de Chazelle worked in the region of Bordeaux according to the best principles of Patullo. They—

[1] Guillory, *Le Marquis de Turbilly, agronome angevin au XVIIIe siècle*, Paris, 1862.

[2] Condorcet, *Eloge of Fougeroux*, pp. 437-8. Arthur Young, *Travels*, vol. 1, pp. 170-1.

[3] Buffon, *Oeuvres Complètes*, Editions Lanessan, Paris, 1884, vol. 1, Notice biographique, p. 19. According to Le Blanc, he lived 'in the heart of France, as people live in England'. See Condorcet, *Eloge de Buffon*, p. 329.

[4] In 1740 a Parisian *bourgeois* founded a Society for the cultivation of rice (*R.H.E.S.* 1928), but it is an isolated example. Nothing happened on a large scale before 1750.

began with the most judicious examination of the soil in every part of their purchase, took the necessary time for making several experiments on it, and gained the best knowledge of its various natures in their power; they published proposals of labourers and all other workmen. . . . . Seventeen hundred farms were established, each consisting of a tract of land from one hundred to one hundred and fifty acres. They placed the habitation, barn, stable, garden and orchard in the centre of each division, one half of it was converted into arable lands and the other into pasture and wood; besides all this, the undertakers furnished the farmers with all the cattle necessary for their first works. . . .[1]

As the edicts promised many material advantages to foreigners undertaking *défrichements*, we find a few English cultivators who took it up. In Calais, 'l'exemple de deux brasseurs anglais qui ont acheté quelques arpents de terre où ils ont établi la culture anglaise dont ils tirent les plus grands profits, déssille les yeux de leurs voisins et contribue plus que tous les livres et mémoires à répandre la bonne culture.'[2] The example was followed by a tradesman who bought land and had it cultivated by English peasants. Was not Arthur Young himself tempted to settle in Bourbonnais?[3]

The *défrichements* sometimes led to a direct modification of the agrarian structure to conform with the best English principles. La Galaizière, Chancellor of Lorraine, applied some of the new agricultural principles which were in favour at the Court of Luneville. He decided to carry out the grouping of tenures on his estates. He also decided on the partition of common lands in these *seigneuries*. The inhabitants expressed their opinions and gave their consent. The operation was registered by the Council in

[1] A. Young, *Letters concerning the present state of the French nation*. Also in *Mercure de France*, August, 1762, p. 107.

[2] 'Observations faites par M. de Bacalan, Intendant du Commerce 1768', *R.H.D.E.S.* 1908, pp. 378 ff.

[3] *Travels*, vol. I, p. 378.
Settlement of English farmers was certainly common enough. The Delportes had English labourers and agricultural specialists on their estate. Members of the nobility had English grooms for their horses. Some actually employed Englishmen to farm their estates. See Blaikie, *Diary of a Scotch gardener at the French Court*, London, 1931, pp. 114, 151, 158, 171, 188.

1771 and Marc Bloch said that it looked like an English bill of enclosure.[1]

Under the reign of Louis XVI, even when the enthusiasm for *défrichements* had faded away, various signs still remained of a certain amount of interest for their estates among the nobility. Fashion and anglomania drove members of the aristocracy into their estates for the hunting season.[2] *Les Liaisons Dangereuses*, or Talleyrand's *Mémoires*, tell us of country life in these *châteaux* where people were beginning to spend a few weeks in the year. Estates were thus sometimes improved. Often the garden alone benefited from the *propriétaire's* solicitude and from his taste for English gardening.

However, independently of this rather artificial return to the country, there is plentiful evidence to show that the *seigneurs* had a definite interest in the welfare of their estate. The Count of Montlosier for instance, was the typical *gentilhomme campagnard*, sincerely interested in agriculture. In his remote province of Auvergne he read books on husbandry (which he did not always like) and experimented in stock-breeding, cultivation of clover, irrigation of meadows.[3]

Exiled Ministers tried to console themselves for the loss of their Parisian life. Choiseul worked much on his Chanteloup estate and built a model sheep-pen.[4] The Counts of Maupeou and Maurepas undertook agrarian reforms on their lands. The Count of Laura-

---

[1] M. Bloch, *A.H.E.S. loc. cit.* 1930, p. 536. About the changes performed at Roville see R. Dion, *op. cit.* Dion thinks that the idea was suggested to La Galaizière by the alterations in the division of the lands (after F. de Neufchâteau) at Rouvres, performed in 1697. But certainly the influence of the new theories is also noticeable in the *Lettres royales* of 1771, which authorized the new grouping. The arguments are the same as those of the anglophile agronomes and expressed in almost similar terms.

[2] According to Carré (*op. cit.*), p. 77, anglomania was noticeable in hunting. The Duc d'Aiguillon in 1768 ordered 18 hounds in England; see A. Young, *Travels*, vol. I, p. 162.   [3] Montlosier, *Mémoires*, Paris, 1830, vol. I, pp. 65 ff.

[4] G. Maugras, *La disgrâce du duc et de la duchesse de Choiseul*, Paris, 1903, p. 120. Also see p. 221.

guais, a notorious anglophile, was exiled in 1778 on his Picard estate of Manicamp. There, 'il fit de l'agronomie, acheta des chevaux, des bœufs, des moutons, et entreprit des expériences'. The enterprise proved ruinous and two years later he owed Lord Tattersall £40,000 for purchases made in England.[1] Voltaire himself at Ferney boasted that he was a practical agronome and scornfully dismissed the agricultural literature. He wrote, 'Experientia rerum magistra'.[2]

Apart from this somewhat forced interest, others, genuinely devoted to the cause, made changes on a large scale. The district of Meillant in Berry, property of the Duc de Béthune-Charost, was revived by the philanthropist *grand seigneur*. He founded there an agricultural Society to which he gave a bore of fifty feet, in order to encourage the study of soils. Thanks to his devotion to agriculture, the district was enriched by artificial grasses, new ploughs, aerated stacks and so on. Before the Crown itself gave the example, he freed his peasants from every form of feudal servitude.[3] This spirit of liberalism was followed in the Provincial Assemblies and we read that the tendency prevailed in the Assembly of Haute-Guyenne (1780) where the *seigneurs agronomes* decided to relinquish their feudal rights.[4]

These may be isolated cases. But they are sufficiently numerous to indicate a liberal tendency and the awakening of a genuine interest in agriculture among the Nobility. The fact that these gentlemen propagated new crops and new instruments shows that the propaganda of the agronomes had been fruitful. If it did not effect any radical changes in the agrarian structure, it certainly led to improvements.

Be this as it may, the last ten years of the period produced

[1] P. Fromageot, 'Le Comte de Lauraguais', *Revue des Etudes Historiques*, 1914.
[2] Voltaire to Moreau de la Rochette, Ferney, 18 January, 1768, in *Mémoires d'Agriculture*, vol. IV, p. 288.
[3] 'Notice sur le duc de Béthune-Charost', *Mémoires d'Agriculture*, An. x, vol. III, p. 338.　　　　　　　　　　　　　　　　[4] Cliquot-Blervache, *op. cit.*

certain typical achievements of the agronomic movement, in the North of France, where the English influence seems, naturally enough, to have been stronger than anywhere else. So far, the model farm at Liancourt was considered as typical of the estates cultivated à l'anglaise.[1] Yet, at the same time, other estates were following this line in a more striking way. Two of these are particularly worth mention: the first one is Marquis de Lormoy's estates in Marquenterre.[2] Having been granted a royal allotment of land, he decided to cultivate it by the English process. The 'Ferme de Châteauneuf' extended to an area of 1,200 acres plus 500 acres of allotment. The principles of the new husbandry were introduced there together with a flock of English sheep. The animals, bought in England, had been shared between de Lormoy, the Ducs de Coigny and de Liancourt, the Marquis de Conflans,[3] and some were even sent to the Royal pen of Trianon. Cattle on this estate were so beautiful that Mesdames Adelaide and Victoire ordered for their model cow-house of Bellevue some specimens 'de la meilleure et de la plus belle espèce d'Angleterre'.[4] The animals had been previously examined by representatives of the Royal Society of Agriculture who gave the most favourable report.

The other and more interesting estate was that of MM. Delporte near Boulogne. It was also a Royal allotment and Roland had very carefully examined it. The initial stages had been difficult, and reports not always enthusiastic. But in 1788 the Marquis de Guerchy could write, 'Cet établissement qui est porté maintenant à huit cent bêtes élevées à la manière anglaise, est de la plus grande beauté et mérite à tous égards la protection que le Gouvernement lui a déjà accordée.'[5] In 1812 the estate had attained a stable prosperity. Success was slow in coming, but it did come finally.

[1] A. Young, Travels, ed. Sée, vol. III, ch. XXXII.

[2] Mémoires de la Société Royale d'Agriculture, Paris, 1788.

[3] He was a notorious anglomane and bought for the Provincial Assembly of Normandy 100 English rams. In A. Young, Travels, vol. II, p. 782.

[4] Calonne, op. cit.   [5] Instructions sur la manière de soigner les bêtes à laine.

The description made of it at that time shows an estate which could successfully vie with the finest ones in Norfolk.[1]

These two achievements seem as the climax to the intensive activity of applied husbandry that was abroad immediately before the outbreak of the Revolution. But other regions of France also had their model estates. The estate of Lavoisier at Fréchines, on which he had spent £120,000 in fourteen years, was on the way to prosperity in 1793, and was cultivated entirely according to the new method.[2] In Perche, the estate of M. Dussieux was managed in the same way. Potatoes were grown on a large scale, ewes and rams from England were bred there, and it was kept with great care.[3] In the South, the Baron Picot de la Peyrouse followed the the example of Norfolk agriculture.

The controversy on sheep had brought into being several centres of breeding in the kingdom. The model pens created by Daubenton in Burgundy, formerly with the support of l'Averdy and Trudaine, were still subsidized by the Government, and in 1784 proved a brilliant success. His efforts were continued in Berry by the Duc de Béthune-Charost, M. d'Amour, Quatremere d'Isjonval, the Archbishop of Bourges, Phelypaux.[4] The supremacy of this province in wool-growing was becoming firmly established. In 1784 an 'Ecole de Bergerie et de Parcage' according to Daubenton's principles was founded by the Provincial Assembly of Berry.[5] In the Ile de France the Marquis de Guerchy and several others obtained 'des succès bien capable d'encourager'.[6] In Alsace, after a difficult start, MM. de Hell and de Montjoie were completely

---

[1] Though the undertaking was still much criticized in H. de Rosny, *Histoire du Boulonnais*, Amiens, 1873, vol. IV, pp. 272-3.

[2] E. Grimaux, *op. cit.* ch. v, pp. 165-6.

[3] G. Bourgin, 'Statistiques révolutionnaires. Description du Département d'Eure et Loir', *R.H.D.E.S.* 1910, p. 340.

[4] Daubenton, *Instruction pour les bergers*, Introduction.

[5] Daubenton, *Mémoire sur le premier drap. Observations pratiques sur les bêtes à laine dans la Province de Berry, par M. le Chevalier de La Merville, Adioint de l'Administration Provinciale du Berry*, Paris, 1789.

[6] *Instruction pour les bergers*, Introduction.

successful, thanks to the advice they received from Daubenton.[1] The Government then took an important measure. In 1787 the Model Farm and Royal *Bergerie* of Rambouillet were organized under the direction of Tessier.[2] This establishment sheltered the merinos and English sheep which were to regenerate the French breeds. It safely underwent the revolutionary changes and still exists. In the provinces, model sheep-folds were set up by the Administration. They were granted privileges and exemptions and some became considerably important, like that of Bove-les-Amiens, for instance, which gathered 190 native ewes, two Dutch ones, two Adrianople rams, and eight English ewes and rams.[3]

Governmental solicitude was extended to technical establishments. The first one dated from 1771. Bertin was its founder and the agronome Sarcey de Sutières was in charge of it. The school established at Anel near Compiègne taught soil-cultivation chiefly. But it was criticized because Sarcey adhered too closely to the English method.[4] The veterinary schools of Lyon and Alfort were preparing the generation of the great veterinary surgeons of the nineteenth century. Official institutions, like the Agricultural Society of Paris and the Comité d'Agriculture, possessed experimental stations. The Abbé Rozier founded at Lyon an 'Ecole pratique pour l'éducation des arbres forestiers'.[5] At Brunoy, Moreau de la Rochette directed a 'Pépinière royale'.[6] In 1785 an

[1] *B.P.E.* 1787.

[2] 'Notice relative à l'Etablissement d'économie rurale de Rambouillet, par M. Tessier', *Mémoires d'Agriculture*, An. XIII, vol. VIII, pp. 11-39.

[3] Calonne, *op. cit.* p. 141. See Baron de Girardot, *Essai sur les Assemblées Provinciales, 1778-90*, Bourges, 1845, pp. 389-90.

[4] Calonne, *op. cit.* And 'Essai sur la nécessité de faire entrer dans l'Institution publique l'enseignement de l'Agriculture', *Mémoires d'Agriculture*, An. x, vol. IV, p. 56, which says: 'C'est à quoi ne font pas assez d'attention les auteurs ou les traducteurs qui vont sans cesse nous chercher des modèles et des leçons dans la pratique des Anglais.'

[5] *Journal de Physique*, 1789, vol. XXXV, pp. 317 ff. On the Abbé Rozier, see A. Young, *Travels*, vol. I, pp. 128-9.

[6] See *Mémoires d'Agriculture*, An. x, vol. IV, pp. 268 ff.

experimental garden was started in Hyères, under a pupil of Thouin. The one in existence at Lorient under Louis XV was about to be re-organized.[1] Royal protection was given to Parmentier, who successfully achieved his experiments on potato-growing. Taste for botany and gardening spread among the nobility. The Marquis of Turgot, the Marquis de la Galissonière, the Baron de Tschudi were famous for their knowledge on exotic trees. The *amateur de fleurs*, typical of the nineteenth century, appeared then for the first time. The foundation of the celebrated Maison Vilmorin-Andrieux, which specialized in the seed-trade, was a result of this newly-acquired taste. Vilmorin was in close touch with England, from which he obtained all sorts of exotic seeds and trees.[2]

Versailles itself symbolized the new tendency. Even if the model farm of Trianon was but a whim of the Queen, a certain amount of importance was inevitably attached to anything attempted. Eighteenth-century Versailles might well be compared to a beautiful exhibition where visitors might see displayed all the tendencies of the Government. La Quintinie's kitchen-garden, the botanical garden of Trianon, the English farm of Marie-Antoinette —they all embodied certain trends of the *ancien régime*.[3]

What was the fate of these institutions, the outcome of this activity? Many of the models disappeared on their founders' deaths. Some were swept away by the Revolution. Many remained, and were transformed, but at least left some traces of their achievements. From 1809 to 1814, the Agricultural Society of Paris instituted an enquiry on agriculture throughout France.[4] Its

[1] Pigeonneau et de Foville, *op. cit.* pp. 92 and 105.

[2] 'Eloge de Vilmorin', *Mémoires d'Agriculture*, vol. x, p. 310.

[3] 'Rapport fait à la S.A. sur la nécessité de conserver l'établissement rural de l'ancienne ménagerie de Versailles', *Mémoires d'Agriculture*, vol. IX, p. 970; Blaikie (*op. cit.* p. 136) still spoke of the Trianon's garden as 'one of the first Botanick gardens of Europe.'

[4] *Mémoires d'Agriculture*, vol. XIII, pp. 139-45, 145-507; vol. XIV, pp. 288-345, 347-471, 478-512; vol. XVI, pp. 241-56, 333-462; vol. XVII, pp. 33-121.

proceedings contain the results of the improvements introduced in some regions over a period of fifty years. In these valuable documents it can be seen that the work of these pioneers had not been in vain.[1] A huge enterprise, like the modernization of French agriculture could not possibly have been achieved in fifty years. Therefore the period 1750-89 must be considered as the pioneering period of modern French agriculture, when new principles were laid down, new techniques tested. The results were undoubtedly retarded by the conditions of the time. It was a period of great enthusiasms, of great aspirations towards reform, a period extremely lively in every respect. But discouragement was also swift in coming, and above all, a great many problems which had been accumulating for a century, required solution simultaneously. Even had the agronomic movement been more widespread, no sudden change could have been effected in French agriculture. Its development since the Revolution shows this clearly enough. It was liable only to improvement, not to complete transformation. None the less, the example of England was basically of the greatest importance, for it gave to the movement the necessary impetus towards modernization and not only taught new techniques but also showed how to apply them in practice. All the practical farmers mentioned here must not therefore be considered as a mere list of ephemeral names. Each of them made a contribution in his particular region. It is thanks to them that the research of the agronomes was assimilated and translated into fruitful experiment, instead of remaining unused in books.

[1] See H. Sée, 'Les progrès de l'agriculture en France de 1815 à 1848', *R.H.E.S.* 1921, p. 67. He considers after G. Bourgin, 'La Révolution et l'Agriculture', *R.H.D.E.S.* 1911, the improvements under Napoleon as a consequence of measures taken by the Revolutionary Assemblies. It is just, however, to remark that these practical improvements (p. 72) and the picture drawn of agricultural efforts after 1830 are but a transposition of what had been advocated by the agronomes.

CONCLUSION

WHAT was the result, finally, of the movement which the agronomes started, and how far did the English influence affect the
movement itself?

Any attempt to answer this question should not lose sight of
two important points: (*a*) the verdict of the eighteenth century
itself; (*b*) our present historical perspective which somewhat
modifies that verdict.

We find this statement at the close of our period:

> Il nous semble résulter de ce qui a été dit, écrit et fait: Que les livres,
> Mémoires, Essais modernes sur l'agriculture en général et la culture
> particulière des plantes qui font l'objet des travaux du laboureur, du
> vigneron, du jardinier pépiniériste, maraicher, fleuriste, ont rendu les
> connaissances sur les divers produits qu'on peut tirer de la terre et sur
> les moyens dont on y parvient, plus communes qu'elles ne l'étaient
> precédemment dans la classe des gens aisés et riches—Que le goût des
> gens aisés et riches pour raisonner et s'instruire des divers objets
> d'agriculture a fait passer quelques cultures usitées dans certains cantons
> dans d'autre qui ne s'en occupaient point, spécialement les pommes de
> terre, les gros navets, turnips ou rabioules.[1]

Two facts are particularly worth noticing in this moderate and
penetrating statement. First, the new acquisitions began to spread;
secondly, taste and knowledge for the new method of farming
were limited to a certain class of society, namely, the more
well-to-do.

In regard to modern conclusions that in 1789 French husbandry
was not much transformed, the value of the movement initiated
by the agronomes may, on the face of it, seem negligible, and has

---

[1] *B.P.E.* 1789, vol. I, pp. xvi, xvii.

in fact sometimes been dismissed.[1] Figures and statistics are not generally concerned with changing agricultural conceptions, which in their turn have little bearing on these symbols of material and immediate results.[2]

It has been the purpose of this study to show that 'possible' or 'potential' facts, 'virtualities' also have their special importance.[3] They did not immediately affect the state of agriculture. Yet in 1789 the new theory, not yet 50 years old, had opened all the avenues that led to our modern conceptions.

Were these advocated changes in the practices of centuries an inevitable tendency? The argument has often been presented in connection with the general mechanism of economic cycles or general economic trends of the century.[4]

It is a matter of conjecture, however, whether the economic conditions which led English agriculture into new ways[5] had the same effect on France,[6] or whether the introduction of the new husbandry in the latter country had its direct origin in these statistical facts.

It has often been pointed out that some great English land-owners, compelled by economic pressure, followed the new depar-

[1] See: E. Levasseur, 'Des progrès de l'agriculture française dans la seconde moitié du XVIIIe siècle', *Revue d'économie politique*, 1898; H. Sée, 'Les origines économiques et sociales de la Révolution française', *R.H.E.S.* 1934-5, pp. 360-74.

[2] Labrousse, *La crise de l'Economie française à la fin de l'Ancien Régime et au début de la Révolution*, Paris, 1944.

[3] This was the attitude of the agronomes towards French farming. Writing about A. Young, De Pradt (*op. cit.*) says, 'A travers les riches moissons dont son sol est couvert [France] il a su entrevoir celles dont sa culture la prive et qu'une autre culture y ferait naître', p. xxi. He writes very soundly on the same subject in ch. II, 'Progrès de la culture en France depuis cent ans,' pp. 13-35, especially p. 33. [4] Labrousse, *op. cit.* pp. 177 ff.

[5] Ernle, Moffit, *op. cit.* An enlightening summary in N. S. B. Gras, *op. cit.* ('Causes of the Agricultural Revolution', and 'Why the Revolution began in England').

[6] Marc Bloch, *Les Caractères originaux*, p. 221. We differ slightly from his theory in attempting to give the English influence a special importance which helped in France 'la grande vague de fond, la Révolution agricole'.

tures in their agricultural methods, and forced them on the peasants. N. S. B. Gras depicts well the mentality of these 'distinguished amateurs', who possessed even then all the psychological traits of capitalists, whereas in France it was only theorists and scientists[1] who enthusiastically embraced the new venture.

De Fresne's books explained very lucidly how French agriculture was transformed[2] in spite of itself, and indeed[3] while following the backward conceptions of its tradition. In 1789 the impossibility of regenerating French farming, or rather of increasing its productiveness by means of the old system, was obvious. For in the struggle between a modernized conception of farming and the maintenance of tradition, the former had not materially won the competition.

The final measurable result was therefore that of routine, and it is too often the only one which has been measured at all. It would therefore be only fair, in viewing eighteenth-century agriculture in France, to attempt to weigh this continuous activity in the field of agricultural research.

Furthermore, the English influence on these new conceptions appears, even more than the conceptions themselves, as a force in its own right, dissociated from the general economic mechanism, especially when we remember that the progress of French farming in modernization was relatively slow in the nineteenth century. The obstacles raised by the feudal system were, therefore, perhaps less important than is commonly believed. Not only did the traditionalist mentality of the French peasant oppose a

[1] Those which the *Nouveau Cours d'Agriculture*, p. xxii, calls 'l'Agronome spéculateur'.

[2] *Op. cit.* Also in A. Young, *Letters concerning the present state of the French nation*, Dublin, 1769, pp. 37, 38.

[3] There certainly was an increase in areas given to cultivation: H. Sée, *Histoire économique*, vol. I, p. 204. See very enlightening remarks by Labrousse (*op. cit.* Preface) which show how an agriculture whose primary object was the production of grain was threatened by the general economic evolution of the period, especially at the end of the *ancien régime*.

successful barrier to it, but also geographical conditions peculiar to France raised one even more insurmountable.[1]

That is why it may be considered that the insistence of the agronomes on English methods is less a fact of economic history, than a fact of the history of ideas.[2] The role of individuals, of public opinion, of fashion thus set, remains important, as opposed to general economic trends, in which however it is included. This may explain why the success of the movement was limited in practice, while very great in theory.

[1] See R. Dion, *op. cit.*

[2] G. Bourgin wrote in 'L'Agriculture, la classe paysanne et la Révolution française', *R.H.D.E.S.* 1911, vol. IV, p. 156: 'Ce qui caractérise, à mon sens, le XVIIIe siècle au point de vue agricole, c'est la disproportion qui existe entre les programmes et les espérances d'une part, les résultats de l'autre.'

APPENDIX I

W H E R E A S official reports were not always favourable in regard to Royal concessions of land granted to M. M. Delporte in Boulonnais and the Marquis de Lormoy in Ponthieu (see Pigeonneau et de Foville, *op. cit.* pp. 336, 337) it might not be without interest to present other views about these estates which offer an excellent instance of cultivation *à l'anglaise*.

Besides Roland de la Platière's 'Etat du troupeau du sieur Delporte, de sa manufacture de tricot, et réflexions sur sa méthode et ses projets', in *Journal de Physique*, 1779, vol. XIV, pp. 93 ff., we find vivid descriptions in Marquis de Guerchy's 'Notions sur l'agriculture de plusieurs cantons de Picardie' in *Bibliothèque Physico-Economique* (1789):

Le plus beau domaine de ce canton est la ferme de Châteauneuf, appartenant actuellement à M. de Lormois qui en a fait l'année dernière l'acquisition et qui est dans l'intention, l'année prochaine, de le faire valoir lui-même: c'est une bien belle entreprise s'il réussit, car cette ferme contient 1200 arpents. Le fermier actuel a 100 chevaux y compris les juments poulinières, 150 vaches ou génisses et 1000 moutons; les vaches sont d'une fort petite espèce: on les dit fort bonnes laitières; les moutons sont à peu près de la même espèce que ceux de Picardie.

M. de Lormois a fait venir, l'année dernière, pour commencer son établissement, 4 béliers et 90 brebis anglaises de la plus belle espèce, les béliers pesant 140 à 150 livres chacun à la tonte. J'ai vu des agneaux de six mois qui pesaient de 100 à 104 livres; leur laine est longue mais de belle qualité: malgré cela il est à craindre que des animaux qui ont été poussés de graisse tout l'été, ne dépérissent dans l'hiver et ne soient trop lourds pour la génération: plusieurs fermiers des environs ont envoyé des brebis au parc de M. de Lormois pour y être servis par ses béliers anglais. . . .

L'établissement de M. de Lormois pourra devenir très intéressant s'il

219

poursuit la culture avec soin, et s'il meuble le domaine, comme il le projette, tant en belles vaches et taureaux d'Irlande, qu'en moutons anglais tirés du Romney Marsh et du Lincolnshire.

Le plaisir que m'a fait cet établissement naissant m'a engagé à en aller voir un beaucoup plus avancé, et qui, dans un an, sera à sa perfection ; celui de M. M. Delporte près de Boulogne ; il est moins étendu que le précédent ne contenant que 400 arpents, mais la culture et les élèves y sont déjà poussées fort loin. C'est un enclos d'une seule pièce, qui jadis était un bois mal venant et plein de broussailles qu'ils ont entièrement défriché ; ils ont bâti une fort jolie habitation sur une hauteur, à une extrêmité du domaine, mais d'où on découvre la totalité. Cette maison, de la plus grande simplicité, est bâtie à la manière anglaise ainsi que tout ce qui est alentour ; la ferme, située à quelques pas de là est construite dans les mêmes principes ; point de granges ; tous les grains et fourrages dont ils ont une très grande quantité, sont en meules, mais très grandes, ayant jusqu'à 100 pieds de long ; il n'y a que deux granges, d'une travée chacune, pour battre chaque espèce de grains.

Les vacheries sont faites aussi comme en Angleterre, avec une cour vis à vis, où les bestiaux vont et viennent à volonté et se roulent sur le fumier.

Le plus beau bâtiment de cette ferme est la bergerie ; c'est un grand carré d'un arpent entouré de chaque côté de 28 travées d'hangars tous ouverts par devant, sous lesquels les brebis peuvent manger à deux rangs de râteliers. Le troupeau n'y entre que quatre mois de l'année, étant le reste du temps au parc ; deux rangs de claies en croix divisent la bergerie en quatre enclos.

Dans le premier sont les béliers ; dans le deuxième les brebis pleines ; dans le troisième les antenoises que l'on ne veut pas qui le deviennent ; enfin dans le dernier les moutons à l'engrais. La totalité de ce troupeau, qui est tout de race anglaise de la plus belle espèce à laine courte est de 5 à 600 et doit être portée à 1,000 suivant le traité du propriétaire avec le Roi, lors de la concession du terrain ; ce troupeau se plait singuliérement dans cet endroit où il est toujours en plein air. M. M. Delporte, pour prouver au Gouvernement la bonté des laines qu'il en retire, en fait fabriquer à Boulogne des tricots aussi beaux que ceux qui sont venus d'Angleterre.

Toute leur ferme est cultivée à la manière anglaise ; ils ont toujours un tiers de leurs terres en trèfle, luzerne ou sainfoin ; ils ont aussi pour leurs vaches deux enclos en ray-grass qui ont trés bien réussi, il en a été de même de la petite minette dorée dont ils ont fait l'essai; ils ont fait avec beaucoup de succès en grand la culture des pommes de terre et des turnips; ils avaient cet été vingt arpents de ces derniers de la plus grande beauté; ils ne les font point arracher et les moutons les mangent dans des claies placées à cet effet. . . .

Finally, in 1812, this model estate had a well-established reputation (*Mémoires d'Agriculture*, vol. xv, pp. 359 ff.):

A force de travaux, de soins, de peines et de facilités pécuniaires, le domaine de Pernes est depuis longtemps un des mieux cultivés, et par conséquent l'un des plus productifs du canton de Boulogne. Il ne nous est pas possible d'offrir dans cet article la balance des dépense et des bénéfices résultant de ce défrichement et de la culture qui en fut la suite . . . cependant nous avons lieu de croire que les résultats d'abord peu avantageux, sont devenus très importants depuis l'augmentation des troupeaux à laine fine et l'éducation des chevaux et du gros bétail, par l'abondance des engrais, le choix des semences, l'adoption de bons assolements, la suppression totale des jachères, et enfin par tout ce qui contribue à donner de la valeur aux terres.

APPENDIX II

THE agricultural revolution in England had sometimes been very well understood by some agronomes. De Fresne, in his *Traité d'Agriculture*, exposes very penetrating views when comparing it with that of France.

We give here the notes with which he explains this comparative evolution and the summary of his conclusions as they appear in the *Bibliothèque Physico-Economique* of 1790.

The cost of English husbandry is diminished

(1) A raison de la double étendue des pâturages.

(2) Dans le rapport du double produit obtenu par leur amélioration.

(3) Ils ont été relativement moindres de moitié, dans le rapport du double produit de chaque arpent de terre labourée.

(4) Dans celui de la consommation des pailles, qui ne servaient que de litière dans l'ancienne disposition de l'égalité, et qui ont augmenté de moitié l'avantage des autres fourrages que les Anglais se sont procurés.

(5) A raison de l'économie que procurent les clôtures en permettant de laisser les bestiaux dans les champs toute l'année, ce qui épargne une prodigieuse quantité de fourrages secs, concourt avec les autres avantages à en diminuer le prix et évite des constructions coûteuses et le transport des fumiers.

(6) A raison de l'économie et de l'avantage que les Anglais se sont procurés en ne nourrissant les menus bestiaux que d'herbes courtes que les gros ne peuvent consommer.

(7) Dans le rapport de la prodigieuse augmentation de bestiaux, dont la destruction procure 6, 9, 16 et jusqu'à 32 fois plus de subsistances, de matières premières, d'engrais, de dépouilles et de travaux de fabrication, relativement à leur consommation, que les bœufs de culture que nous avons multipliés.

(8) A raison de l'économie de temps que l'on obtient, et de la moitié du terrain que l'on gagne, en servant de chevaux au lieu de bœufs pour les labours.

PLATE II

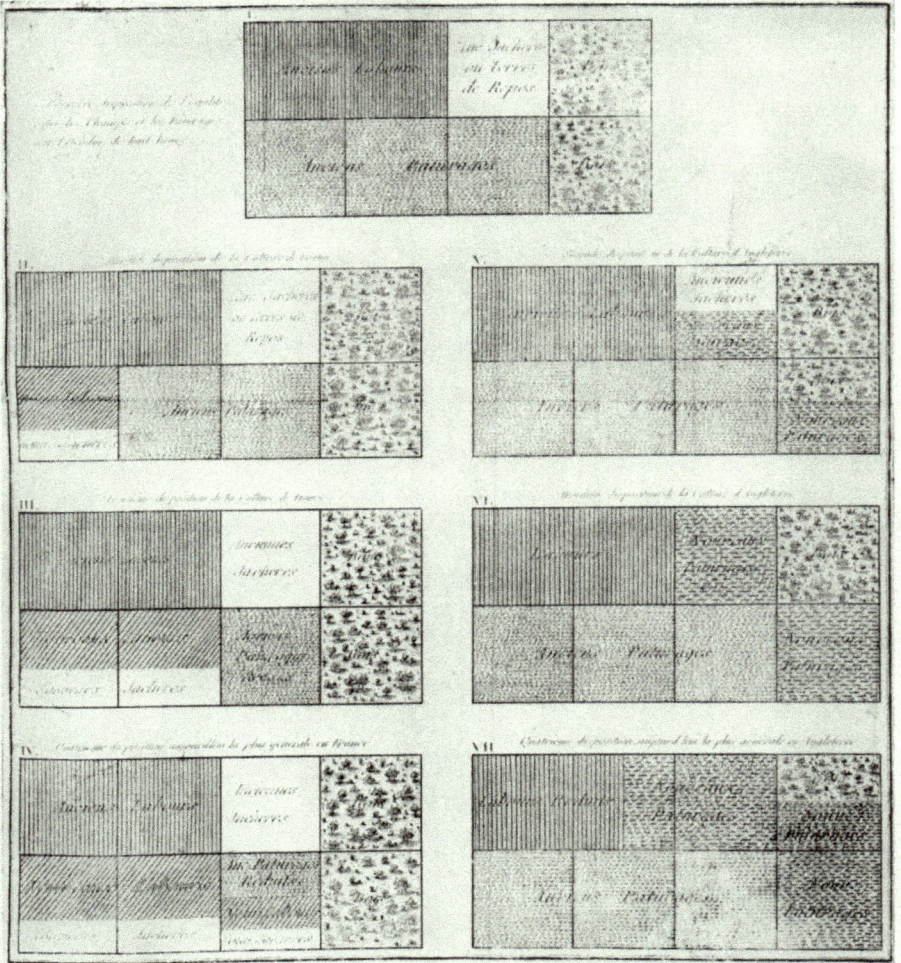

COMPARAISON D'AGRICULTURE DE FRANCE ET D'ANGLETERRE
(*Bibliothèque Physico-économique, Année* 1790)

   I. Première disposition de l'égalité entre les champs et les pâturages sur l'étendue de huit lieues.
  II. Seconde disposition de la culture de France.
 III. Troisième disposition de la culture de France.
 IV. Quatrième disposition aujourd'hui la plus générale en France.
  V. Seconde disposition de la culture d'Angleterre.
 VI. Troisième disposition de la culture d'Angleterre.
VII. Quatrième disposition aujourd'hui la plus générale en Angleterre.

(9) A raison de l'économie que procurent les clôtures, soit en augmentant l'effet des engrais, soit en garantissant les grains du dommage des bestiaux et les bestiaux de la voracité des loups, soit par les autres avantages qu'on a expliqués.

(10) A raison de l'économie que les Anglais se sont procuré, en ne construisant aucuns bâtiments pour leur fourrages, ni même pour leurs grains; ils les entassent le plus souvent en gerbes dans leurs champs ou dans leurs cours.

(11) Dans le rapport de la réduction d'un quart des labours.

(12) Enfin à raison de l'avantage que procurent les canaux et la facilité des chemins et des labours, en rendant à la culture une prodigieuse quantité de fourrages.

BIBLIOGRAPHY

Only books having a definite connection with the subject have been mentioned here. It seemed unnecessary to give too lengthy a list of works, however useful they have proved for the elaboration of this study, but from which no particular details have been retained.

I. MODERN STUDIES

A. GENERAL STUDIES, BIBLIOGRAPHICAL SOURCES, BIO-GRAPHIES, ETC.

A list of current periodicals (*Economic History Review, Journal of Modern History, Revue d' Histoire Moderne*, etc.) is not given here. It should be pointed out, however, that *Agricultural History* published quarterly since 1928 contains several valuable studies on eighteenth-century English farming, especially by G. E. Fussell (references given in footnotes).

Biographie Universelle (Michaud), 2nd ed. 45 vol. 1842-65.

F. Brunot, *Histoire de la langue française des origines à 1900*, vol. VI, Paris, 1930.

Cambridge Modern History, vol. VI, Cambridge, 1909.

A. Cherest, *La chute de l'Ancien Régime*, Paris, 1884, 3 vol.

Dictionary of National Biography, London, 1885-1901, 63 vol.

F. Funck-Brentano, *L'Ancien Régime en France*, 1926.

A. Garrigou-Lagrange, *Production agricole et économie rurale*, Paris, 1939.

P. Gaxotte, *Le siècle de Louis XV*, Paris, 1933.

Gide et Rist, *Histoire des doctrines économiques depuis les physiocrates*, Paris, 1909.

La Grande Encyclopédie, 1885-1901, 31 vol.

N. S. B. Gras, *A History of Agriculture in Europe and America*, New York, 1925.

H. Higgs, *Bibliography of Economics, 1751-75*, Cambridge, 1935.

J. B. Huzard, *Catalogue de la Bibliothèque de J. B. Huzard*, Paris, 1842, 3 vol.

J. Jaurès, *Histoire socialiste de la Révolution Française*, Paris, 1922.

Jobez, *La France sous Louis XVI*, Paris, 1877-93, 3 vol.

E. Lavisse, *Histoire de France*, vols. VIII, IX, Paris, 1909.

Lavisse et Rambaud, *Histoire générale*, vol. VII, Paris, 1896.

G. Lefebvre, 'La Révolution Française', *Peuples et civilisations*, vol. XIII, 1941.

E. Lipson, *The Economic History of England*, London, 1931, 3 vol.

G. Lizerand, *Le Régime rural de l'ancienne France*, Paris, 1942.

M. Marion, *Histoire financière de la France*, vol. I, *1715-89*, Paris, 1913.

—— *Dictionnaire des Institutions de la France aux XVIIe et XVIIIe siècles*, Paris, 1923.

Germain Martin, 'Histoire économique et financière de la France', *Histoire de la Nation Française dirigée par G. Hanotaux*, Paris, 1927.

Montague-Fordham, *A Short History of English Rural Life*, London, 1918.

Pierre Muret, 'La prépondérance anglaise', *Peuples et civilisations*, vol. XI, Paris, 1937.

G. Pagès, *La Monarchie française d'Ancien Régime*, Paris, 1928.

Rothamsted Experimental Station, *Library catalogue*, Second edition, 1940.

P. Sagnac, 'La fin de l'Ancien Régime et la Révolution Américaine', *Peuples et Civilisations*, vol. XII, Paris, 1941.

Scottish peerage, Edinburgh 1904-14, 9 vol.

H. Sée, *Esquisse d'une histoire du régime agraire en Europe aux XVIIIe et XIXe siècles*, Paris, 1921.

—— *Histoire économique de la France* (with notes by R. Schnerb), Paris, 1939.

Svenskt Biografiskt Lexikon, vol. I, Stockholm, 1918.

B. SPECIAL STUDIES

Alem, *Le marquis d'Argenson et l'économie politique*, Paris, 1900.

E. Allix, 'Propriété foncière et fortune mobilière sous la Révolution', *R.H.E.S.* 1913.

Annales de l'Agriculture française. Paris, An. IV.

P. Ardascheff, 'Les Intendants de province à la fin de l'Ancien Régime', *Revue d'Histoire moderne et contemporaine*, vol. V, 1903.

Ascoli, *La Grande Bretagne devant l'opinion française au XVIIe siècle*, Paris, 1930, 2 vol.

A. Babeau, *La Province sous l'Ancien Régime*, Paris, 1894, 2 vol.

—— *Le village sous l'Ancien Régime*, Paris, 1878.

—— *Les voyageurs en France depuis la Renaissance jusqu'à la Révolution*, Paris, 1885.

Bacquié, *Les inspecteurs des manufactures sous l'ancien régime (1669-1792)*, *Mémoires et documents de J. Hayem*, 11e série, 1927.

Ch. Ballot, *L'introduction du machinisme dans l'industrie française*, Paris, 1923.

—— 'La fondation du Creusot', *R.H.D.E.S.* 1912.

W. Berman, *History and Art of warming and ventilating rooms and buildings*, London, 1845, 2 vol.

Y. Bezard, *La vie rurale dans le sud de la Région Parisienne de 1450 à 1560*, Paris, 1929.

R. Black, *Horse racing in France*, London, 1886.

—— *Horse racing in England*, London, 1893.

M. Bloch, *Les Caractères originaux de l'histoire rurale française*, Oslo, 1931.

—— 'Les plans parcellaires' (1929); 'La lutte pour l'individualisme agraire dans la France du XVIIIe siècle' (1930); 'La vie rurale, jadis et naguère' (1930), *A.H.E.S.*

P. Boissonnade, 'Les débuts de l'industrie cotonnière en France—1760-1806', *R.H.D.E.S.* 1914.

P. M. Bondois, 'La protection du troupeau français au XVIIIe siècle. L'Epizootie de 1763', *R.H.E.S.* 1932.

G. Bonno, *La Constitution Britannique devant l'opinion française de Montesquieu à Bonaparte*, Paris, 1932.

G. Bourgin, 'Le partage des biens communaux', *Collection des documents Economiques de la Révolution*, Paris, 1908.

—— 'Les communaux et la Révolution française'. *Nouvelle revue historique de droit français et étranger*, 1908.

—— 'Statistiques révolutionnaires—Description du département d'Eure et Loir' (1910); L'Agriculture, la classe paysanne et la Révolution française' (1911), *R.H.D.E.S.*

E. R. Briggs, *The Political Academies of France in the Early 18th century, with special reference to the 'Club de l'Entresol' and to its founder, the Abbé Alary*, Cambridge, Ph.D. dissertation.

A. Britsch, *La Jeunesse de Philippe Egalité*, Paris, 1926.

——*Lettres de L. P. J. d'Orléans à N. P. Forth, 1778, 1785*, Paris, 1926.

de Calonne, *La vie agricole sous l'Ancien Régime en Picardie et en Artois*, Paris, 1883. *Ibid.* edition of 1920.

Marquise de Campana de Cavelli, *Les derniers Stuarts et la Cour de St Germain en Laye*, Paris, 1871, 2 vol.

E. Carcassonne, *Montesquieu et le problème de la Constitution britannique au 18e siècle*, Paris, 1927.

P. Caron, 'Etat de l'Agriculture dans la généralité d'Amiens, 1788.' *Bulletin d'histoire économique de la Révolution*, 1909.

H. Carré, *La noblesse de France et l'opinion publique au 18e siècle*, Paris, 1920.

E. Cavailhon, *Les haras de France*, Paris, 1886.

E. Champion, *La France d'après les Cahiers de 1789*, Paris, 1897.

Sir J. Clapham, *The Economic Development of France and Germany 1815-1914*, Cambridge, 1936.

Corneille de Witt, *La société anglaise et la société française au XVIIIe siècle*.

G. Cuvier, *Histoire des Sciences Naturelles*, Paris, 1843.

D. Dakin, *Turgot and the 'ancien régime' in France*, London, 1939.

Dauzat, *Le Sentiment de la Nature et son expression artistique*, Paris, 1914.

Demangeon, *La plaine Picarde*, Paris, 1905.

H. Denis, *Les prohibitions et les entraves à la libre exploitation des terres*. Paris, 1911.

E. Depitre, 'Le Système et la querelle de la Noblesse commerçante, 1756-9,' *R.H.E.S.* 1913.

Comte de Dienne, *Histoire du dessèchement des lacs et marais en France avant 1789*, Paris, 1891.

R. Dion, *Essai sur la formation du paysage rural français*, Tours, 1934.

H. Doniol, *Histoire des classes rurales en France*, Paris, 1857.

F. Dreyfus, *La Rochefoucault-Liancourt*, Paris, 1903.

Comtesse Drohojowska, *Les grands agriculteurs modernes*, Tours, 1885.

L. Dutil, *L'état économique du Languedoc à la fin de l'Ancien Régime*, Paris, 1911.

Lord Ernle, *English farming past and present* (5th ed.), London, 1936.

—— *The pleasant land of France*, London, 1908.

G. Fagniez, *L'économie sociale de la France sous Henry IV, 1589-1610*, Paris, 1897.

B. Fay, *La Franc-Maçonnerie et la Révolution intellectuelle du XVIIIe siècle*, Paris, 1935.

Octave Festy, *L'Agriculture et la Révolution Française*, Paris, 1947.

G. de la Fournière, 'Les Comités d'Agriculture de 1760 et de 1784'. *Bulletin du Comité des travaux historiques et scientifiques. Section, Des sciences économiques et sociales*, 1909.

E. Freffe, *Histoire des troupes étrangères au service de France*, Paris, 1854. 2 vol.

P. Fromageot, 'Le Comte de Lauraguais', *Revue des études historiques*, 1914.

G. E. Fussell, *The Old English Farming Books*, London, 1947.

—— *From Tolpuddle to T.U.C.*, London, 1948. (Contains an important bibliography of the contributions to post-mediaeval English agricultural history.)

F. C. Green, *Minuet*, London, 1939.

E. Grimaux, *Lavoisier*, Paris, 1888.

Guillory, *Le marquis de Turbilly, agronome angevin au XVIIIe siècle*, Paris, 1862.

J. L. and Barbara Hammond, *The Village Labourer, 1760-1832*, London, 1919.

A. S. Haslam, *The Biography of Arthur Young*, Rugby, 1930.

Paul Hazard, *La crise de la conscience européenne (1680-1715)*, Paris, 1935.

——*La pensée européenne de Montesquieu à Lessing*, Paris, 1945, 3 vol.

H. Higgs, *The Physiocrats*, London, 1899.

Ch. Hoffman, *L'Alsace au XVIIIe siècle*, 1906, 2 vol.

Kareiew, *Les paysans et la question paysanne en France dans le dernier quart du XVIIIe siècle*, Paris, 1899.

K. Kautsky, *La question agraire, étude sur les tendances de l'agriculture moderne*, Paris, 1900.

L. Knowles, 'Economic causes of the French Revolution', *The Economic Journal*, 1919.

M. Kowalewsky, *La France économique à la veille de la Révolution*, Paris, 1909, 2 vol.

Krug-Basse, *L'Alsace avant 1789, ou état des institutions provinciales et locales*, Paris, 1877.

Labiche, *Les Sociétés d'Agriculture au XVIIIe siècle*, 1908.

Labrousse, *La crise de l'Economie française à la fin de l'Ancien Régime et au début de la Révolution*, Paris, 1944.

Léonce de Lavergne, *La Société d'Agriculture de Paris*, Paris, 1870.

―― *Les Assemblées Provinciales sous Louis XVI*, Paris, 1864.

―― *Economie rurale de la France*, Paris, 1866.

Dom Leclerc, *Histoire de la Régence*, Paris, 1922, 3 vol.

G. Lefebvre, 'Les recherches relatives à la répartition de la propriété et de l'exploitation foncières à la fin de l'Ancien Régime', *Revue d'histoire moderne*, 1928.

―― *Les paysans du Nord pendant la Révolution française*, Lille, 1924.

Levainville, *L'Industrie du fer en France*, Paris, 1922.

Levasseur, 'Des progrès de l'agriculture française dans la seconde moitié du XVIIIe siècle', *Revue d'Economie politique*, 1898.

H. Levy-Bruhl, 'La noblesse de France et le commerce à la fin de l'Ancien Régime', *Revue d'histoire moderne*, Mai-Juillet, 1933.

E. Lipson, *The History of the Woollen and Worsted Industries*, London, 1921.

H. Lockitt, *The Relations of French and English Societies, 1783-93*, London, 1920.

E. G. Lodge, *Sully, Colbert and Turgot*, London, 1931.

de Loménie, *Les Mirabeau*, Paris, 1879, 3 vol.

Loutchisky, *La propriété paysanne en France à la veille de la Révolution*, Paris, 1912.

Baron de Maricourt, *En marge de notre histoire. Un monde disparu: Jardiniers et concierges du Roi*, Paris, 1905.

M. Marion, 'Les classes rurales en Bordelais au XVIIIe siècle', *Revue des études historiques*, vol. IV, 1902.

T. H. Marshall, 'J. Tull and the "New Husbandry"', *Economic History Review*, 1929.

G. Martin, *La Grande Industrie en France sous le règne de Louis XV*, Paris, 1900.

Mathorez, *Les Etrangers en France*, Paris, 1919-21, 2 vol.

G. Maugras, *La disgrâce du duc et de la duchesse de Choiseul*, Paris, 1903.

Mauguin, *Etudes historiques sur l'administration de l'agriculture*, Paris, 1877, 3 vol.

A. Maury, *L'ancienne Académie des Sciences*, Paris, 1864.

C. Maxwell, *The English traveller in France 1750-1815*, London, 1932.

E. Menault, *Souvenirs de Beauce, Cassegrain, Blanchet, Tessier*, Paris, 1859.

L. W. Moffit, *England on the Eve of the Industrial Revolution*, London, 1925.

D. Mornet, *Les sciences de la nature en France au XVIIIe siècle*, Paris, 1911.

—— *Les origines intellectuelles de la Révolution française*, Paris, 1933.

H. Munro-Cassidy, *Un voyageur philosophe au XVIIIe siècle, J. B. Leblanc*, (Harvard Studies in Comparative Literature), Cambridge (Mass.), 1941.

R. Musset, *Histoire de l'élevage du cheval en France*, Paris, 1917.
'L'Administration des haras en France au XVIIIe', *Revue d'histoire moderne et contemporaine*, 1909-10, vol. XIII.

R. Parisot, *Histoire de Lorraine*, Paris, 1922, 2 vol.

D. Pasquet, 'La découverte de l'Angleterre par les Français au 18e siècle', *Revue de Paris*, 1920, 1921.

H. Passy, *Des systèmes de culture et de leur influence sur l'économie rurale*, Paris, 1853.

L. Passy, *Histoire de la Société Nationale d'Agriculture*, Paris, 1914.

G. Prato, 'L'evoluzione agricola nel secolo XVIII', *Memorie della Reale Accademia di Scienze di Torino*, 2nd series, vol. LX, 1910.

E. J. B. Rathery, *Des relations sociales et intellectuelles entre la France et l'Angleterre*, Paris, 1856.

de Rémusat, *L'Angleterre au XVIIIe siècle*, Paris, 1857.

P. Renouvin, *Les Assemblées Provinciales de 1787*, Paris, 1921.

Georgia Robinson, *La Réveillière-Lépeaux, Citizen director 1753-1824* (Columbia University Ph.D. thesis), New York, 1938.

H. de Rosny, *Histoire du Boulonnais*, Amiens, 1873, 5 vol.

M. Rouff, *Les Mines de charbon en France 1774-91*, Paris, 1922.

G. Roupnel, *La Ville et la Campagne au XVIIe siècle. Etude sur les population du pays Dijonnais*, Paris, 1922.

F. Roze, *Histoire de la pomme de terre*, Paris, 1898.

P. Sagnac, *La société française sous l'Ancien Régime*, Paris, 1947. 2 vol.

E. Savoy, *L'Agriculture à travers les âges*, Paris, 1940, 3 vol. (Vol. 3, by R. Grand and R. Delatouche, 'L'Agriculture au Moyen Age de la fin de l'Empire Romain au XVIe siècle'.)

G. Schelle, *Dupont de Nemours et l'école physiocratique*, Paris, 1888.

H. Sée, *Les origines du capitalisme moderne*, Paris, 1926.

—— *Les idées politiques en France au XVIIIe siècle*, Paris, 1920.

—— *La vie économique et les classes sociales en France au XVIIIe siècle*. (Bibliothèque générale des sciences sociales.) Paris, 1924.

—— 'Le partage des biens communaux à la fin de l'Ancien Régime'. *Nouvelle revue historique du droit*, 4e série, vol. II, 1923.

—— 'The economic and social origins of the French Revolution', *Economic History Review*, 1931-2.

—— 'Un type de document, le Livre de raison d'un Parlementaire breton au 18e siècle', *A.H.E.S.* 1931.

—— 'La mise en valeur des terres incultes . . . à la fin de l'Ancien Régime' (1923); 'Les progrès de l'Agriculture en France de 1815 à 1848' (1921), *R.H.E.S.*

J. Sion, *Les Paysans de la Normandie orientale*, Paris, 1909.

F. C. Smith, *The Early History of Veterinary Literature*, London, 1915-23.

H. Taine, *Les origines de la France contemporaine*, vol. I, 'L'Ancien Régime', Paris, 1875.

J. Texte, *J.-J. Rousseau et le cosmopolitisme littéraire*, Paris, 1898.

A. de Tocqueville, *L'Ancien Régime et la Révolution*, Paris, 1860.

A. P. Usher, *A history of mechanical inventions*, New York, 1929.

P. de Vaissière, *Gentilshommes campagnards de l'ancienne France*, Paris, 1903.

—— *Les Curés de Campagne au XVIIIe siècle*, Paris, 1933.

Vermale, *Les classes rurales en Savoie au XVIIIe siècle*, Paris, 1911.

Viallate, *L'activité économique en France de la fin du XVIIIe siècle à nos jours*, Paris, 1937.

E. J. M. Vignon, *Etudes historiques sur l'administration des voies publiques en France aux 17e et 18e siècles*, Paris, 1862, 4 vol.

L. Villat, *La Révolution* (Collection Clio), vol. I, Paris, 1936.

G. Weulersse, *Le mouvement physiocratique en France*, Paris, 1910. 2 vol.

—— 'Le mouvement préphysiocratique en France', *R.H.E.S.* 1931.

F. Wolters, *Agrarzustande und Agrarprobleme in Frankreich von 1700 bis 1790*, Leipzig, 1905.

P. Yvon, *Traits d'union anglo-normands*, Caen, London, 1919.

BIBLIOGRAPHY

II. ORIGINAL SOURCES

A. PERIODICALS

Académie Royale des Sciences. Histoire de l'Académie depuis 1666 (continued as *Mémoires*), Paris, 1666-1790.

L'Année Littéraire, Paris, 1754-70, 122 vol.

Bibliothèque Britannique, La Haye, 1733-47.

Bibliothèque Britannique (edited by Pictet), Genève, 1796-1800.

Bibliothèque physico-économique, 1782-91.

Journal de Physique (Observations sur la physique, sur l'histoire naturelle et sur les arts ... par M. l'Abbé Rozier), Paris, 1773-93, 43 vol.

Journal des Savants et Mémoires de Trévoux, Amsterdam, 1750-81.

Journal Oeconomique, 1751-2-5, etc.

Mémoires d'Agriculture . . . publiés par la Société d'Agriculture du Département de la Seine, Paris, 1801-19, 22 vol.

Mémoires de la Société Royale d'Agriculture de Paris, 1785-9.

Mercure de France, Paris, 1781-91.

Nouvelliste oeconomique et littéraire, La Haye, 1754-8, 22 vol.

Royal Society of London: Philosophical transactions, 1750-90.

B. DICTIONARIES, ENCYCLOPAEDIAS, COLLECTIONS OF DOCUMENTS, ETC.

Actes de l'Académie impériale et royale de Bruxelles, 1792.

L'Agronome, ou Dictionnaire portatif du cultivateur, Paris, 1762.

Annales de Chimie, Paris, 1789-93, 10 vol.

Annales du Muséum, Paris, 1802-6, 5 vol.

N. Chomel, *Dictionnaire oeconomique*, 2 vols, Paris, 1730. 2 supplementary volumes, 1743.

Cours Complet d'Agriculture théorique et pratique . . . par une Société d'Agriculteurs et rédigé par Rozier, Paris, an. IX, 6 vol.

Cours complet d'Agriculture . . . par M. le baron de Morogues, Paris, 1840.

Dictionnaire d'Agriculture . . . par les membres de la section d'Agriculture de l'Institut, Paris, 1809, 13 vol.

Dictionnaire des Sciences Naturelles . . . par plusieurs Professeurs du Jardin du Roi, Strasbourg, Paris, 1816.

Dictionnaire de Trévoux, Paris, 1752. 7 vol.

Dictionnaire universel d'agriculture et de jardinage, Paris, 1751.

Encyclopédie ou Dictionnaire raisonné des sciences, des arts et métiers, 1751-5, 17 vol.

Encyclopédie Méthodique (Section Agriculture), Paris, 1787-1821, 7 vol.

Encyclopédie Méthodique (Art aratoire et du jardinage), 2 vol. 1791.

Abbé d'Expilly, *Dictionnaire*, 6 vol. Paris, 1762-70.

Gazette Littéraire de l'Europe, Paris, 1764-6. 8 vol.

Histoire et Mémoires de l'Académie Royale des Sciences, Inscriptions et Belles Lettres de Toulouse, Toulouse, 1782.

P. Jaubert, *Dictionnaire Universel des Arts et Métiers*, Paris, 1773.

Loudon, *An Encyclopaedia of Agriculture*, London, 1825.

Musset-Pathay, *Bibliographie agronomique*, Paris, 1810.

Nouveau cours complet d'Agriculture théorique et pratique . . . par les membres de la Section d'Agriculture de l'Institut, Paris, 1809.

Abbé de Petity, *Encyclopédie élémentaire*, Paris, 1767, 3 vol.

Abbé Rozier, *Cours Complet d'Agriculture . . . ou Dictionnaire universel d'Agriculture*, Paris, 1781, 9 vol.

Savary des Bruslons, *Dictionnaire universel de commerce*, Copenhague, 1759-65.

Valmont de Bomare, *Dictionnaire raisonné, universel, d'Histoire Naturelle*, Paris, 1768.

C. TREATISES, SPECIAL STUDIES, MISCELLANEOUS

L'Abondance rétablie ou moyens de prévenir en France la disette des bestiaux en même temps qu'on rétablit la fertilité de la terre, Paris, 1769.

C. R. Aikin, *An account of the most important recent discoveries and improvements in Chemistry*, London, 1814.

Amoreux, *Mémoire sur les haies*, Paris, 1787.

Androuet du Cerceau, *Les plus excellents bastiments de France* (1576-9).

Marquis d'Argenson, *Considérations sur le gouvernement ancien et présent de la France*, Amsterdam, 1764.

Arnould, *De la Balance du Commerce*, Paris, an. III.

L'art d'augmenter et de conserver son bien, ou règles générales pour l'administration d'une terre, Paris, 1784.

L'art de battre, écraser, piler, moudre et monder les grains avec de nouvelles machines, Paris, 1769.

'Observations faites par M. de Bacalan, Intendant du Commerce, 1768,' *R.H.D.E.S.* 1908.

Baert, *Tableau de la Gde Bretagne, de l'Irlande et des possessions anglaises.* .. , Paris, An. 8, 3 vol.

Béardé de l'Abbaye, *Essays in agriculture: or a variety of useful hints for its improvement*, translated from the French, 1776.

Bellepierre de Neuve Eglise, *L'Agronomie et l'Industrie, ou les principes de l'agriculture et du commerce*, Paris, 1761, 6 vol.

Bernard, *Mémoire sur les engrais que la Provence peut fournir, et sur la manière de les employer, suivant les diverses espèces de terreins*, Marseille, 1778.

J. Bertrand, *De l'eau relativement à l'économie rustique, ou traité de l'irrigation des prés*, Lyon, 1764.

R. Billing, *An account of the culture of carrots; and their great use in feeding and fattening cattle*, 1765.

Duc de Biron (Lauzun), *Mémoires (1747-83)*, Paris, 1859.

Blaikie, *Diary of a Scotch gardener*, Paris, 1931.

P. Boissonnade, 'Trois mémoires relatifs à l'amélioration des manufactures de France sous l'administration des Trudaine, 1754', *R.H.E.S.* vol. VII, 1914.

Boncerf, *Les inconvénients des droits féodaux*, Londres, 1776.

Le Bon Fermier ou l'Ami des Laboureurs, Paris, 1767.

Boucher d'Argis, *Code rural, ou Maximes et réglements concernant les biens de campagne*, Paris, 1774, 2 vol.

Bourgelat, *Eléments d'Hippiatrique ou Nouveaux principes sur la connaissance et la Médecine des chevaux*, Lyon, 1751.

Boussole agronomique ou Guide des Laboureurs ... traduit du Latin par quatre curés de Normandie, Paris, 1762.

R. Bradley, *Observations physiques et pratiques sur le jardinage et l'art de planter, avec le calendrier des jardiniers, ouvrage traduit de l'anglais.* .. . Paris, 1764.

—— *Le calendrier des jardiniers*, Paris, 1743.

G. Brice, *Nouvelle description de la Ville de Paris*, Paris, 1725.

P. J. Buc'hoz, *Histoire universelle du règne végétal.* .. , Paris, 1775-80, 13 vol.

Buffon, *Oeuvres complètes* (édition Lanessan), Paris, 1884, 14 vol.

—— *Correspondance inédite* (édition Nadault de Buffon), Paris, 1860, 2 vol.

—— *Mémoire sur la conservation et le rétablissement des forêts*, 1739.

—— *Mémoire sur la culture des forêts*, 1742.

—— *La statique des végétaux et l'analyse de l'air* (from S. Hales), Paris, 1735.

Butel-Dumont, *Recherches historiques sur l'administration des terres*, Paris, 1779.

—— *Essai sur le commerce d'Angleterre*, Paris, 1755.

Butré, 'De la grande et de la petite culture', *Ephémérides du citoyen*, 1767, vol. IX-XII.

W. Camden, *Britannia; or a chorographical description of Great Britain and Ireland, together with the adjacent islands*, translated from the Latin by Edmund Gibson, 1722, 2 vol. (1st edition 1586).

Abbé Carlier, 'Mémoire sur les moyens de perfectionner les laines de la France'; 'Observations historiques sur l'état ancien et l'état actuel des troupeaux et des laines en Angleterre'. *Journal de Physique*, 1784.

—— *Mémoire sur les laines*, Bruxelles, 1755.

—— *Considérations sur les moyens de rétablir en France des bonnes espèces des bêtes à laine*, Paris, 1762.

—— *Instructions sur la manière d'élever et de perfectionner la bonne espèce des bêtes à laine de Flandre*, 1763.

—— *Traité des bêtes à laine*, Compiègne, 1770, 2 vol.

Cels et F. H. Gilbert, *Instructions sur les effets des inondations et débordements des rivières, relativement aux prairies, aux récoltes de foin et à la nourriture des animaux*, Paris, 1795.

Chantreau, *Voyage dans les trois Royaumes d'Angleterre . . . fait en 1788 et 1789*, Paris, 1792.

Chaptal, *Elements of Chemistry*, London, 1791, 3 vol.

—— *De l'Industrie Française*, Paris, 1819, 2 vol.

A. Chatillon, *Topographie Française. Représentation des plusieurs villes, bourgs, plans. . .* , 1648.

Cliquot-Blervache, *Essai sur les moyens d'améliorer en France la condition des Laboureurs, des Journaliers . . . par un Savoyard*, Chambéry, 1789.

De Commerell, *Mémoire sur l'amélioration de l'agriculture par la suppression des jachères*, Paris, 1798.

Condorcet, *Oeuvres* (Edition O'Connor et Arago), Paris, 1847, 12 vol.

Condorcet et Turgot, *Correspondance inédite*, Paris, 1882.

Corps d'observations de la Société d' Agriculture de Bretagne, 1757-8

Costa de Beauregard, *Essai sur l'amélioration de l'agriculture dans les pays montueux et en particulier dans la Savoie*, Chambéry, 1774.

Abbé Coyer, *La Noblesse commerçante*, London, 1756.

—— *Développement et défense du système de la Noblesse commerçante*, Amsterdam, 1757.

Cretté de Palluel, *Traité des prairies artificielles*, Paris, 1801.

Daire, *Les Economistes français*, Paris, 1846.

Dangeul, *Remarques sur les avantages et les désavantages de la France et de la Grande Bretagne . . . traduites de l'Anglais du Chevalier John Nickolls*, Amsterdam, 1754.

Daubenton, *Instruction pour les bergers et les propriétaires des troupeaux* (Introduction by Huzard), 1800.

—— *Mémoire sur le premier Drap de Laine superfine du cru de la France*, Paris, 1784. *Addition au mémoire précédent*, 1784.

—— *Observations sur la comparaison de la nouvelle laine superfine de France*, 1785.

Sir Humphrey Davy, *Elements of Agricultural Chemistry*, London, 1827.

Delaguette, *Culture du pêcher*, 1750.

Delisle, *Mémoire sur le ray-grass et le Red Clover*, 1761.

De Lisle de Froncel, *Méthodes et projets pour parvenir à la destruction des Loups dans le Royaume*, Paris, 1768.

L. B. Desplaces, *Préservatif contre l'Agromanie, ou l'Agriculture réduite à ses vrais principes*, Paris, 1762.

—— *Histoire de l'Agriculture ancienne*, Paris, 1765.

M. Despommiers, *L'art de s'enrichir promptement par l'Agriculture prouvé par des expériences*, Paris, 1769.

Kenelm Digby, *A Discourse concerning the Vegetation of Plants*, 1669.

Duhamel du Monceau, *Traité de la Conservation des Grains et en particulier du Froment*, Paris, 1753.

—— *La physique des arbres*, Paris, 1758.

—— *Des semis et plantations d'arbres*, Paris, 1760.

—— *Traité des arbres et des arbustes qui se cultivent en France en pleine terre*, Paris, 1755, 2 vol.

—— *Eléments d'Agriculture*, Paris, 1763, 2 vol.

—— *Traité de la Culture des Terres suivant les principes de M. Tull, Anglais*, Paris, 1750-6, 6 vol.

Dumont de Courset, *Mémoire sur l'Agriculture du Boulonnais et des cantons maritimes voisins*, Boulogne, 1784.

Dupont de Nemours, *De l'exportation et de l'importation des grains*, Paris, 1764.

Dupuy-Demportes, *Le Gentilhomme maréchal, de l'Anglais de J. Barthelet*, Paris, 1756-8, 2 vol.

—— *Le Gentilhomme Cultivateur tiré de l'Anglais et de tous les auteurs qui le mieux écrit sur cet art*, Paris, 1761, 3 vol.

Ecole d'Agriculture, Paris, 1759.

W. Ellis, *The timber tree improved; or the best practical methods of improving different lands with proper timber*, 1738.

Les Loisirs du chevalier d'Eon, Amsterdam, 1775, 13 vol.

J. D'Epremesnil, *Correspondance sur une question politique d'agriculture*, Amsterdam, Paris, 1763.

L'Espion Chinois, Cologne, 1744, 6 vol.

Essai sur l'Etat de la Culture Belgique et sur les moyens de la perfectionner, London, 1784.

Essais de la Société de Dublin, Traduits de l'Anglais par Dr Thébault, Paris, 1759.

Comte d'Essuile, *Traité des Communes . . . où joignant la politique à l'économie, on démontre leur inutilité, le préjudice qu'elles font à l'agriculture et l'avantage que l'on retirerait de leur aliénation et de leur partage*, Paris, 1770.

Fabroni, *Réflexions sur l'état actuel de l'Agriculture*, Paris, 1780.

Faujas de St Fond, *A Journey through England and Scotland* (edition with notes and a memoir of the author by Sir A. Geikie), Glasgow, 1907, 2 vol.

Forbonnais, *Les Eléments du Commerce*, Paris, 1754.

Foreign essays on agriculture and arts, 1765, 1766.

Fourcroy, *Système des connaissances chimiques et de leurs applications aux phénomènes de la nature et de l'art*, Paris, an. IX, 10 vol.

——*Leçons élémentaires d'Histoire naturelle et de chimie*, Paris, 1782, 2 vol.

Fresnais de Beaumont, *La Noblesse Cultivatrice*, Paris, 1778.

De Fresne, *Traité d'Agriculture*, Paris, 1788, 3 vol.

de Fréville, *Voyage agronomique, précédé du parfait fermier, ouvrage traduit de l'Anglais*, Paris, 1774.

Abbé Froger, *Instructions de Morale, d'Agriculture et d'Economie . . . ouvrage déstiné à servir pour enseigner à lire aux enfants de la Campagne*, Paris, 1759.

Gerbaux et Schmidt, *Procès verbaux des Comités d'Agriculture et de Commerce de la Constituante*, Paris, 1906.

Gilbert, *Traité des Prairies artificielles*, Paris, 1790.

—— *Recherches sur les espèces de prairies artificielles qu'on cultive avec le plus d'avantage en France*, Metz, 1801.

M. W. Gilpin, *Voyages en differentes parties de l'Angleterre, traduit de l'Anglais par M. Guidon de Berchère*, Paris-Londres, 1789.

A. Goudar, *Les intérêts de la France mal entendus dans les branches de l'Agriculture . . . par un citoyen*, Amsterdam, 1756.

Goyon de la Plombanie, *La France agricole et marchande*, Avignon (Paris), 1762.

Baron Grimm, *Correspondance littéraire, philosophique. . .* , 1877-87, 20 vol.

J. B. Grosley, *Londres*, Lausanne (Paris), 1774, 4 vol.

Marquis de Guerchy, *Mémoire pour l'amélioration des bêtes à laine dans l'Ile de France*, Paris, 1785.

—— *Instructions sur la manière de soigner les bêtes à laine suivant les principes de M. Daubenton*, n.d.

de la Guérinière, *Ecole de cavalerie contenant la connaissance, l'instruction et la conservation du cheval*, Paris, 1761.

Guyton de Morveau, *Eléments de chymie théorique et pratique*, Dijon, 1777, 3 vol.

Th. Hale, *A compleat body of husbandry*, London, 1758, 4 vol.

F. Hastfer, *Instruction sur la manière d'élever et de perfectionner les bêtes à laine, traduite du Suédois*, Paris, 1756.

W. Henry, *An epitome of chemistry*, London, 1803.

John Hill, *Eden or a Compleat body of gardening*, London, 1757.

Histoire générale des voyages, Paris, 1761, 16 vol.

F. Home, *The principles of agriculture and vegetation*, Edinburgh, 1757.

—— *Les principes de l'agriculture et de la végétation*, Paris, 1761.

Instruction sur la culture des Turneps ou gros navets, Imprimée par ordre du Roi, 1786.

Instruction sur le parcage des bêtes à laine, Publiée par ordre du Gouvernement, Paris, 1785.

Lacretelle, *Histoire de France*, Paris, 1809, 8 vol.

E. G. Lafosse, *Observations et découvertes faites sur les chevaux. . .* , Paris, 1754.

Lalande, *Des Canaux de Navigation*, Paris, 1778.

De la Marre, *Défense de Plusieurs Ouvrages sur l'Agriculture*, Paris, 1765.

La Merville, *Observations pratiques sur les bêtes à laine, dans la province de Berry*, Paris, 1789.

La Quintinie, *Instruction pour les jardins fruitiers et potagers . . .* , Amsterdam, 1692.

—— *The compleat Gard'ner*, translation by J. Evelyn, London, 1693.

Vie du duc de la Rochefoucauld-Liancourt par son fils, Paris, 1827.

F. de la Rochefoucauld (1765-1848), *La vie en Angleterre au XVIIIe siècle*, Paris, 1945.

La Salle de l'Etang, *Prairies artificielles ou Lettres a M. de . . .* , Paris, 1756.

—— *Manuel d'Agriculture pour le Laboureur, pour le Propriétaire et pour le Gouvernement . . . avec la réfutation de la nouvellemétho de de M. Thull*, Paris, 1764.

Lasteyrie, *Traité des constructions rurales . . . ouvrage publié par le Bureau d'Agriculture de Londres et traduit de l'anglais avec des additions*, Paris, an. x.

Lavoisier, *Oeuvres*, vol. v, 1864-93, 6 vol.

Abbé Le Blanc, *Lettres d'un Français sur les Anglais*, Paris, 1745.

—— *Letters on the English and French Nations*, London, 1747, 2 vol.

Le Blanc, *Recueil de machines, instruments et appareils qui servent à l'économie rurale et industrielle*, n.d. (between 1815-30).

Le Boucher de Crosne, *Mémoire sur les haras*, Paris, Lacombe, 1770.

Lerouge, *Jardins anglais et chinois*, 1776-89.

Le Trosne, *De l'Administration Provincial*, Bâle, 1788.

L'Héritier, *Sertum Anglicum, seu Plantae rariores quae in hortis juxta Londinium, imprimis in horto Regio Kewensi excoluntur*, Paris, 1788.

J. Liebig, *Traité de chimie organique*, Paris, 1840.

L. Liger, *Oeconomie générale de la Campagne ou nouvelle maison rustique*, Paris, 1701, 2 vol.

——*La nouvelle maison rustique*, Paris, 1721, 2 vol.

Linguet, *Annales politiques, civiles et littéraires*, Londres, 1777.

Marquis de Lormoy, *Mémoire sur l'Agriculture*, Paris, 1789.

Lullin de Châteauvieux, *Mémoire sur la pratique du semoir*, Lyon, 1761.

Macquer, *Dictionnaire de chymie*, Paris, 1778, 4 vol.

De Mante, *Traité des prairies artificielles, des enclos, et de l'éducation des moutons de race anglaise*, Paris, 1778.

Manuel des champs. . . , Paris, 1780.

W. Marshall, *The rural economy of Norfolk, comprising the management of landed estates and the present practise of husbandry in the county,* 1787.

Massac, 'Mémoire sur la qualité et l'emploi des engrais', *Journal d'Agriculture,* Juillet-Août, 1767.

Mémoires de la Société d'Agriculture de Dublin, traduits de l'Anglais par M. Thebault, Paris, 1759.

L. S. Mercier, *L'An deux mille quatre cent quarante,* Londres, 1771.

(Ph.) Miller, *Dictionnaire des plantes,* 1735.

—— *The gardener's dictionary,* 1733-9, 2 vol.

Marquis de Mirabeau, *L'Ami des Hommes,* La Haye, 1759, 6 vol.

Dom Miroudot, *Prairies artificielles, Mémoire sur le Fromental et la Culture anglaise, Mémoire abrégé sur le sainfoin ou l'Esparcette tiré des recueils de la Société oeconomique de Berne, auxquels on a joint quelques remarques de Tull et Duhamel* . . . , Lyon, 1762.

Monbron, *Préservatif contre l'Anglomanie,* Paris, 1756.

Comte de Montlosier, *Mémoires,* Paris, 1830, 2 vol.

Abbé Morellet, *Lettres à Lord Shelburne,* Paris, 1898.

Moreri, *Dictionnaire,* ed. 1674.

Mortimer, *Agriculture complète* . . . *ou art d'améliorer les terres,* 1765.

Chevalier Mustel, *Mémoire sur la pomme de terre et sur le pain économique lu à la S.R.A. de Rouen,* 1758.

Le nouveau Newkastle ou nouveau traité de cavalerie, Paris, 1747.

Noel et Carpentier, *Nouveau Dictionnaire des origines, inventions et découvertes dans les arts, les sciences,* Paris, 1803.

Abbés Nolin et Blavet, *Essai sur l'Agriculture moderne,* Paris, 1755.

Baron Ogilvy, *Mémoire sur les semoirs,* 1761.

Parmentier, *Traité sur la culture et les usages des pommes de terre, de la patate et du topinambour,* Paris, 1789.

H. Patullo, *Essai sur l'amélioration des terres,* 1765.

—— *An Essay upon the Cultivation of the Lands and Improvements of the Revenues of Bengal,* London, 1772.

Pecquet, *Loix forestière de France,* Paris, 1753.

Pelé de St Maurice, *L'art de cultiver les peupliers d'Italie,* Paris, 1767.

Picot, Baron de la Peyrouse, *The agriculture of a district in the South of France,* translated from the French, 1819.

Pigeonneau et de Foville, *L'administration de l'agriculture au Contrôle Général des Finances*, Paris, 1882.

R. de Planazu, *Oeuvres d'Agriculture*, Paris, 1778.

Abbé Pluche, *Le spectacle de la Nature, ou entretiens sur les particularités de l'Histoire naturelle*, Paris, 1754-5, 8 vol.

Practical Treatise of Husbandry by the celebrated M. Duhamel du Monceau. London, 1759.

De Pradt, *De l'état de la Culture en France*, Paris, 1802, 2 vol.

Premier essai d'agronomie, ou diététique générale des végétaux, et application de la chimie à l'agriculture. . . , Dijon, 1777.

Abbé Prevost, *Mémoires et aventures d'un homme de qualité*, La Haye, 1757.

Marquis de Puismarais, 'Mémoire sur les moyens de bonnifier les laines. . . ', *Nouvelliste oeconomique*, 1754.

Reboul, *Discours sur les moyens d'encourager l'Agriculture en Provence*, Aix, 1770.

B. Rocque, *A practical treatise of cultivating lucerne*, 1761.

Roland de la Platière, 'Mémoire sur l'éducation des troupeaux et de la culture des laines', *Journal de Physique*, 1779.

Abbé Rozier, *Nouvelle table des articles contenus dans les volumes de l'Académie royale des Sciences de Paris, 1666-1770*, Paris, 1775.

Saint Lambert, *Les Saisons, poème traduit de l'Anglais*, Dordrecht, 1769.

Baron de St Supplix, *Le Consolateur, pour servir de réponse à la théorie de l'impôt*, Paris, 1763.

Sarcey de Sutières, *Agriculture expérimentale*, Paris, 1765.

Abbé Roger Schabol, *La pratique du jardinage*, Paris, 1770.

—— 'Sur les villages de Montreuill, Bagnolet etc.', *Nouvelliste oeconomique*, 1755.

Comte de Ségur, *Mémoires*, Paris, Londres, 1825.

Sénac de Meilhan, *Du gouvernement, des mœurs et des conditions en France avant la Révolution*, Hambourg, 1795.

Olivier de Serres, *Théâtre d'agriculture*, edition of François de Neufchâteau, 1804.

J. de Solleysel, *Le parfait maréchal*, Paris, 1775.

Sonini de Manoncourt, *Mémoire sur la culture et les avantages du chounavet de Laponie*, Paris, 1788.

Abbé Soumille, *L'usage du semoir*, 1755.

—— *Description du semoir à bras de Languedoc*, Avignon, 1762.

Tessier, *Traité des maladies des grains*, Paris, 1783.

Thierriat, *Instructions familières ... sur la culture des terres*, Paris, 1764.

Thomas (maître jardinier du Lord évêque de Lincoln), *Almanack de poche pour cultiver les légumes*, 1776.

Th. Thompson, *History of the Royal Society from its Institution to the end of the eighteenth century*, London, 1812.

M. Tillet, *Essai sur la cause qui corrompt et noircit les grains dans les épis*, Bordeaux, 1755.

—— *Précis des expériences faites à Trianon sur la cause qui corrompt les blés*, Paris, 1756.

—— *Histoire d'un insecte qui dévore les grains dans l'Angoumois*, Paris, 1763.

Traité sur la Culture des Jardins, par le Jardinier de Milord R. Manners, Londres, 1755.

Transactions philosophiques de la Société Royale de Londres, 1731-6, traduites par M. de Brémond, Paris, 1738-41.

Baron de Tschudi, *Traité des arbres résineux conifères, traduit de l'Anglais de Miller*, Metz, 1768.

Jethro Tull, *Horse-hoeing husbandry or an Essay on the principles of tillage and vegetation*, London, 1733.

Marquis de Turbilly, *Mémoire sur les Défrichements*, Paris, 1760.

—— *A discourse on the cultivation of waste and barren lands*, London, 1762.

Abbé de Vallemont, *Curiosités de la Nature et l'Art sur la Végétation, l'Agriculture et le Jardinage dans leur perfection*, Paris, 1703.

Abbé de Véri, *Journal* (édition du Baron Jehan de Witte). Paris, n.d. 2 vol.

Voltaire, *Letters concerning the English nation*, London, 1733.

Arthur Young, *Letters concerning the present state of the French nation*, Dublin, 1769.

—— *A six months tour through the north of England*, 1770, 4 vol.

—— *A six weeks tour through the southern counties of England*, 1778.

—— *Travels in France during the years 1787, 1788, 1789* (ed. Betham-Edwards), London, 1905.

—— *Voyages en France (1787, 1788, 1789)*, Edition H. Sée, Paris, 1931, 3 vol.

INDEX

For EU product safety concerns, contact us at Calle de José Abascal, 56–1°,
28003 Madrid, Spain or eugpsr@cambridge.org.

www.ingramcontent.com/pod-product-compliance
Ingram Content Group UK Ltd.
Pitfield, Milton Keynes, MK11 3LW, UK
UKHW010851090126
466816UK00011B/170